Sustainable Biotechnology

SUSTAINABLE BIOTECHNOLOGY

Editors
Prof. Sudhir U. Meshram
and
Dr. G. B. Shinde

DISCOVERY PUBLISHING HOUSE
NEW DELHI

Published by:

DISCOVERY PUBLISHING HOUSE PVT. LTD.
4383/4B, Ansari Road, Darya Ganj
New Delhi-110 002 (India)
Phone : +91-11-23279245; 23253475; 43596065
E-mail : discoverybooksindia@gmail.com
discoverypublishinghouse@gmail.com
namitwasan9@gmail.com
web : www.discoverypublishinggroup.com

***First Published:* 2003**
***Reprinted:* 2022**

ISBN: 978-81-7141-741-4

Sustainable Biotechnology

Printed at:
Infinity Imaging Systems
Delhi

FOREWORD

Modern biotechnology at the dawn of the new century is beginning to unfold its promise with some spectacular discoveries of great practical importance. The completion of the Human Genome Project announced recently offers enormous possibilities in the field of medicine with the development of new drugs, vaccines and diagnostics. In the field of agriculture the emphasis has been on the production of transgenic varieties of crop plants. Thus, it has been possible to undo millions of years of evolution and transfer genes across completely unrelated species. A particularly, striking example is that of the transfer of beta-carotene genes into the rice plant, so that it can become a source of vitamin-A for millions of malnourished children. Important as these contributions are, the real promise of biotechnology for agriculture is even greater. The hope is that the new genetic manipulations will make it possible to develop a new kind of agriculture, which will be both more productive and more efficient. We are beginning to reach yield ceilings in some of our most important crops like wheat and rice. In the next 25 years it may be possible to increase photosynthetic efficiency in order to overcome these yield barriers. Also, it should be possible to replace the non-renewable resources of energy with those of a renewable kind such as the biologically fixed nitrogen. Scientist will be working to transfer the nitrogen-fixing genes across a wide range of crop plants. Similarly, animal nutrition will be improved with the bioconversion of crop residues to increase their digestibility and nutritive value.

All this will call for new research priorities in the field of biotechnology research. While the private sector will continue to have an important role, it is clear the public sector research will have to be greatly strengthened for taking up some of the long term research programmes to make future agriculture highly productive, efficient and environment friendly. I hope this International Conference will address some of these issues.

Dr. H.K. Jain
Former Dy. Director General
ISNAR
The Hague
Holland

In Present Context

Rising concentration of greenhouse gases in the atmosphere is threatening the world with global rise in its mean temperature which is expected to cause extensive changes in precipitation patterns and conditions of environment. These changes pose a greater threat to plant species as a whole. In order to stay in favourable climate, large-scale migrations shall occur with changes in climate. Species not capable of migration shall be adversely affected and the entire spectrum of life forms on this planet could acquire different character. It is difficult to accurately predict as to how the biosphere will react to the climatic changes triggered by global warming. Many species could become extinct, many others could survive only in reduced numbers, while some may flourish with a changed geographical distribution. Since different species shall respond to the changes in climate patterns in different ways, the entire biotic spectrum over the globe is expected to change. An inevitable consequence of this upheaval could be a drastic reduction in biological diversity.

The United Nations Conference on Human Environment convened in 1972 adopted a resolution concerning International Programme for the preservation of genetic resources of plants cultivated in the tropical regions of the world. Almost simultaneously a global strategy for preservation of seeds was developed at Beltsville, U.S.A. The International Board for Plant Genetic Resources was framed in 1974 with Rome as its headquarters. By 1985, it had set up a chain of 43 gene banks all over the world of which 21 are situated in the developing countries. The seed banks or gene banks or germ-plasm banks earlier used simple techniques for preservation which involved storage of dried seeds under ambient or lower temperatures. General types of seeds (eg., of pulses, cereals, fruits and vegetables) possess natural dormancy and can be easily preserved using short-term, or long-term storage techniques, which involved drying and preserving in sealed containers in temperature ranges of +5° to –20°C.

Some of the seeds of tropical plants, bushes or fruit crops do not possess a natural dormancy. Hence they die quickly if not allowed to germinate immediately. The simple techniques available at that time could not preserve them. They had to be constantly cultivated. Similarly, a number of plants which reproduce by means of their vegetative parts such as tubers, corms, rhizomes etc. had to be regularly cultivated and multiplied.

However, recent techniques of preservation greatly extended the scope of germpalasm banks. Under liquid nitrogen temperature to about –196°C, seeds dried to a low moisture content can be preserved for almost indefinite periods (eg., several centuries). By applying tissue culture and cryo-preservation techniques, almost any type of plant can be preserved in viable state for long durations. Plants which cannot be preserved by traditional methods are preserved in the form of tissues or meristem cultures under liquid nitrogen. Small pieces of simple vegetative tissues or better still few meristematic cells of plants to be preserved are introduced in synthetic medium under asceptic conditions and are allowed to grow into small clumps, called the callus. These cultures or calluses are maintained under liquid nitrogen for any length of time. They can be multiplied and young plants differentiated from the tissues as and when required.

Obviously, there is a dire need to establish several germplasm banks to preserve and conserve local biological diversity in various regions of the developing countries. Further, R&D efforts should continue in developing simpler and more economical techniques for preservation of the valuable biological species.

Urban areas are an antithesis of natural diversity. Only resilient species like weeds and pest thrive in these places. The constant expansion of resources areas, its massive consumption of rebowices and emission of wastes threaten both agriculture and wild life. Wild life conservation has to be achieved by a three-pronged strategy, Viz., controlling ever expanding urban settlements, reducing the damage to the ecosystems and infusing biological diversity into urban settlements.

A major excitement in the new millennium is the great stride made by the scientists involved in the "Human Genome Project" in mapping 97 per cent of human genome and the assurance to complete the rest by 2003. This has tremendous potential to empower the human race which so far has been far from the realm of man. However, it is the subline responsibility of scientists to ensure that the advances in science and technology should not be allowed to take the ethical norms to ransom, and thereby rendering the very scientific progress counter-productive.

Biotechnology, with its immense potentialities and applications in aquaculture, agriculture, medical and immunology, forestry, pollution control and chemical production is one of the few rays of hope to make our planet earth, a better place to live. Any advanced technology is often accompanied by benefits to the mankind, but at the same time, some inherent risks are unavoidable. Therefore, for the efficacious and safe utilisation of biotechnology which is a powerful multi-disciplinary field of far reaching consequences, it is absolutely essential to exercise caution in identifying safe and sustainable applications for the benefit of mankind and for the survival of the planet earth along with its ecological integrity and biological diversity. The scientists involved in research and development in biotechnology should also ensure that their discoveries are solely for the benefit of mankind and its sustainable progress without causing undesirable social and ethical imbalance. All the R&D efforts in this area should be directed towards this objective. Encouragement of indigenous technologies originating from individual innovations is absolutely essential by providing adequate resource and moral support from international and national funding agencies, international bodies and the respective Government agencies.

21st century belongs to Biotechnology and let us all pledge to use its immense potentialities for the benefit of mankind and sustainable development of our society with due regards to our planet earth and all its ecosystems.

Sudhir U. Meshram
Chairman
Organising Committee
Global Sustainable Biotech Congress 2000 AD

CONTENTS

CHAPTER

1

BIOTECHNOLOGY REVOLUTION FROM MICROBE TO MAN

D.P.S. Verma

We are currently witnessing a biotechnology revolution that started about 10 years ago and is expected to continue for the next 10 years. During this period not only the genomes of major crops, microbes and man will be completely mapped but also their functionality determined which is likely to have a far reaching impact on biology and the society at-large. The seeds of this technology were planted even before the discovery of the structure of DNA, but during the last two decades recombinant DNA technology has matured and has begun to show fruits in both medical and agricultural area. Over 70% of agricultural crops in USA are genetically engineered to day and new traits are rapidly being developed. The application of genomics and bioinformatics is expected to further accelerate the use of biotechnology applications for human health, agriculture and industrial uses. This is witnessed by enormous amount of cash infusion of over $19 billion in this sector this year alone. At the same time novel technologies are being developed for identifying new genes, mutations and several new fields of investigations have evolved such as pharmacogenomics, toxicogenomics etc. How India can make

Professor, Molecular Genetics Biotechnology Centre, Ohio State University, Colombus OH 43210 USA.

use of this volcanic eruption of knowledge, technology and opportunities in the field of biotechnology. This can only be done by a change in attitude, approach and willingness to harvest this vast resource. Unfortunately, the progress in India in this direction has been very slow despite the fact that this is the only country in activity. The overall curricula in biology in most Indian universities and colleges has not changed to keep pace with the rapid progress that has occurred in biology. One way India may make a use of this technology is by developing bioinformatics tools to mine the data and develop applications. This would require rapid integration of computer programming in biology. This avenue must be explored as it is most cost effective and beneficial to the society at-large. Finally, entrepreneurial spirit must be developed and nurtured in order to take full advantage of this emerging technology.

CHAPTER

2

BIOTECHNOLOGY AND THE PRINCIPLES OF EFFICACY, SAFETY AND PRECAUTION

Prof. Carlos M. Romeo-Casabona, Ph.D., M.D.*

1. The role of law with regard to Biotechnology research

Gene biotechnology is a powerful instrument in the fight against hereditary diseases, as well as those of microbial origin (viruses, bacteria, fungi, parasites, etc.) and others caused by imbalances in the organism's biochemical functioning. Several drugs already include proteins as an active principle and large quantities of vaccines and diagnostic products for human use have been obtained, firstly from natural organisms and later through genetic engineering, from isolated human genes (e.g. proteins which are lacking in some people, such as insulin, interferon, growth hormone, erythropoietin, factor VIII, tissue plasminogen activator, etc.)

From time immemorial humans have endeavoured to modify living species, both plants and animals, in some way or other, to harness the results for their own use, particularly in farming and stockbreeding. More recently, efforts have centred on industry and particularly pre-clinical research (e.g. the creation of oncomice for use in cancer research) and for therapeutic purposes, as sources of organs and tissues for

Director, Inter University in Law and the Human Genome Appar todo de correos, Post Box-48080, Bilbao-Spain.

humans (for example, modifying certain animals by manipuilating their genes and even inserting human genes), i.e. transgenic animals.

As the foregoing examples show, Biotechnology involves the application of knowledge from numerous areas of science and engineering in order to develop productive processes. Put another way, the application of different techniques to living matter. Today's Biotechnology is based largely on genetic engineering, using the recombinant DNA technique (rDNA). Of particular interest to us here is the one which involves the direct or indirect use of human genes. This does not mean that we should overlook the growing worldwide importance of all the different Biotechnology market sectors, including health, as mentioned above, which is the most highly developed one so far, industrial supplies; food industry (transgenic products); and others affecting more basic facets, such as energy development and the environment. However, the development of such products and the underlying research have to be compatible with the adoption of precautions and safety measures governing the handling of living matter, especially if it is modified genetically and when the interferences in other living creatures, including humans, are unpredictable. The Law, for its part, is finding it is being put on the spot as regards the protection of research results, especially with new products involving living matter, which throw up new facets that cannot always be assimilated readily by traditional legal instruments, all the more so when ethical question marks stand against economic interests that should, in principle, have to be considered legitimate. Let us recall again that, legally-speaking, attention tends to focus more on issues arising from aspects of knowledge of our social reality that give rise to different valuations, particularly where there is opposition to the effects or consequences. Usually these issues turn into inter-individual and social conflicts that may warrant intervention of the Law. It is equally true, however, that this hypothetical intervention is increasingly preceded by appropriate ethical reflection aimed at finding more or less generally-acceptable responses and solutions, addressed primarily to the authorities. This ethical reflection, perceived more and more as indispensable in the case of the Life Sciences generally, should not be ignored therefore in the case of Biotechnology.

Why should the Law intervene also in Biotechnology? The following are some of the most important reasons:

1. To guarantee respect for freedom of thought and freedom of scientific enquiry and creation. As is known, these freedoms have been extended to such a degree that they are now considered basic rights, and thus legal systems must provide the appropriate legal resources to guarantee and protect the exercise of said rights.

2. To promote research in science and technology. This is a goal where, although self-evident, awareness does not always exist of the contribution to be made by legal instruments also, albeit partially and to a limited extent.

3. It is equally true that legal instruments can serve the interests, needs and concerns of the authorities, as specific expressions of their R+D policy actions, particularly those relating to the different aspects of Biotechnology (health, industrial supplies, the food industry, energy and environment, etc), in accordance with the scientific and technological characteristics of the country, its infrastructure and resources, priority R+D needs etc.

4. Alongside the Law's role regarding the promotion of research in science and technology, one would also have to include the protection of research results and Biotechnology innovations, and ultimately the inventions arising therefrom. However, there is no disputing that, as a consequence of this promotion and protection, a social function must be derived also. In other words, the necessary mechanisms should be used to ensure that the population at large can share in the benefits obtained or derived from biotechnologies.

5. To channel the research process itself, in order to ensure the quality, efficacy and safety both of the research and of any ensuing applications. Similarly, and as another necessary effect, to eliminate, prevent, or at least minimize the risks to humans, the environment and above all to living matter generally. In the case of biotechnologies, such concerns are exemplified mainly, although by no means exclusively, in the precautionary principle, whereby recourse to such technologies should be in tandem with advances in science, but only if there are similar advances in our knowledge of the risks and of means of prevention and control, where needed. The precautionary principle, which is grounded on scientific uncertainty and is a recent manifestation of the law of responsibility, implies the adoption of protection measures beyond what is strictly necessary based on the calculation of probabilities of risk deemed unlikely. Furthermore, the notion of permitted risk also serves as a pointer as to how one should behave in connection with certain risk activities which are nonetheless lawful. It helps determine a risk limit which must not be overstepped if civil or criminal liability is to be avoided, depending on the case.

6. Lastly, there may also be another step or level of intervention, which would involve limiting or even prohibiting some potential applications, to the extent that they entail deviations that are clearly harmful to individuals or society. In this specific context, we should recall the problems raised by the use of human gametes, embryos and foetuses for health and pharmacological research and experimentation, as well as the use of animals and living matter generally for such research. Secondly, the discussion on whether restrictions or prohibition should be imposed on research whose aims are most definitely harmful for humans (for example, production of biological weapons or attempts to demonstrate the

existence of different physical or mental attributes among ethnic groups and then rank these groups accordingly).

Having referred to the promotion and protection of research, but also to the possibility that some applications or effects might be directed, limited and even prohibited in some extreme cases, it is necessary to underline some of the many benefits that come with Biotechnology. First and foremost, evidently, are the benefits derived from Biotechnology research and applications. However, a way must be found to safeguard other social values or to harmonise and strike a balance with them. Examples include the environment – in which the general public is taking an ever-growing interest, and biological variety, due to its inherent richness and the need to ensure the survival of all living species. Mention is even made also of the need to safeguard the identity of living matter and more specifically in the case of the human genome, its integrity.

Normatively-speaking, developments have taken place extremely rapidly, thanks mainly to the intense activity at European Union level, for the most part in the form of Directives. Similarly, several countries have enacted their own specific laws on the different implications of Biotechnology and more generally, on the human genome and by extension the potential biotechnology applications resulting from research in this particular field. Germany, Britain, Austria, France, United States and Brazil, to mention but a few, are sufficiently illustrative examples of legislative initiatives on Biotechnology during the 1990s. For its part, Spain has done likewise and the country's lawmakers have been equally sensitive to some of the major issues, at least as regards some sectorial aspects of Biotechnology, as will be shown below. At the same time, it is likely that both the relations between the Administration and the promoters of biotechnology research and also Spain's R+D Programme should render possible a specific legal framework to facilitate a favourable climate for research and investment (particularly to attract venture capital and the setting up of spin-off and start up companies). Such as employment and fiscal legislation (for instance, a range of tax incentives until results or profits are achieved, which usually takes some time), and greater flexibility and speed in the case of the administrative procedures needed to create firms in the biotechnology sector.

2. Safety and efficacy in Biotechnology: controlling the risks of the productive process

With its highly precise and efficacious techniques modern Biotechnology is opening up new and highly promising prospects in various areas of industry, agriculture, food, health etc. The capacity to interfere with living matter is thus much greater. At the same time, however, some of the effects are unpredictable and hence not always controllable. Concern focuses particularly on the risk of alteration of biodiversity and of the ecosystem balance, the modification of the natural evolution of species, thus endangering the survival of some living creatures and organisms, which might directly affect humans. To date, recombinant DNA technology has proven to be safe, thanks largely to the care with which it has been used, unlike what has happened with other better-known technologies.

Much attention has focused on genetic manipulation of microorganisms and their subsequent use. These activities, which are crucial for some production aspects of industry and farming, have had to be limited and regulated by specific legal instruments, tailored to the characteristics of the productive processes. The risks of uncontrolled dissemination of such organisms have shown clearly that bio-safety is an issue that transcends national borders and, as such, may even require certain limitations on sovereignty. For this reason, it has been stressed also that isolated regulations in a country would be futile if neighbouring countries do not adopt similar measures, but rather become genetic havens, or more accurately in this case, Biotechnology havens, especially for multinationals, whose versatility and power enable them to set up where fewest controls and prevention measures are in place, thus guaranteeing a greater return on production. Only if coordinated legislation exists can Biotechnology products be placed on the market under equal conditions.

One result of this new sensitivity have been the initiatives by the Organisation for Economic Cooperation and Development (OECD). In 1986 the Organisation published some considerations on safety in rDNA manipulation, stating that evaluation and management of the risks associated with the obtaining and handling of genetically modified organisms (GMO) using these techniques should be similar to what is already done for other, better-known organisms.

For its part, the European Union, in its 4th Environment Action Programme (1987-1992), concluded that Community action in the new biotechnologies should focus on optimizing their use, to prevent contamination of the environment by evaluating the potential risk and through regulation of two major risk activities, namely, the confined use and deliberate release of microorganisms. As a conclusion to this process, the Council approved two Directives on 23 April 1990: 90/219/EEC on the confined use of genetically modified organisms, and 90/220/EEC on the deliberate release into the environment of genetically modified organisms. Both Directives have since been updated, firstly 1994 and then again in 1997, and further changes are currently being made. They have also produced a ruling by the European Court of Justice. Most EU Member States have transposed the Directives into domestic law. Similar legislative initiatives have also been taken outside Europe. There is no denying that together these represent that first major supranational initiative on activities posing a high risk both to the environment and human health, and one can but regret that the same was not done in the case of nuclear energy.

It is worth noting also that the regulatory approach taken marks a departure in several ways. There is a shift away from control and regulation of a dangerous products and the proposed use thereof (vertical approach, taken by the United States) towards control and regulation of a specific and dangerous technique or activity (horizontal cover, in the EU). A shift away from risk considered ex post to one viewed ex ante. In other words, action is taken with respect to a method or process before the danger actually manifests itself. Under this approach the dangerousness of the process is presumed to exist until evidence to the contrary emerges, a view which is now given preference over the previous one which assumed that production and trade could take place freely as long as no evidence of risk was shown. Put another way, the burden of proof is reversed with regard to the absence of risk in the process. In terms of regulations, the approach has resulted not just in rules governing the activity itself but also the establishment of a series of authorizations and other requirements prior to commencement. As was mentioned elsewhere, these derive from the precautionary principle, which is why I consider Europe's approach and methodology to be more appropriate since they are safer and more efficient in terms of prevention, but at the same time do not impede research or production.

An example of this type of European regulation can be seen in Spain's Law on the subject, which is a transposition of the aforementioned European Directives. In its Statement of Purpose the Law says that it seeks to protect human health, both as regards health care and also pharmaceutical products. It seeks also to prevent risks that might affect the different elements and components of the environment and to regulate research in science and *technology* and the *instruments* used therein.

The Law establishes regulations governing the confined use and deliberate release of genetically modified organisms for research and development purposes or any other purposes outside the provisions laid down in health or pharmaceutical products legislation, and also governing commercial activities involving such organisms or products in which they are present. One could say, therefore, that the regulatory scope of Spin's legislation is wider than the EU version on which it is based (the Directives referred to above). Whereas the latter are confined to microorganisms but also multi-cel living beings, which would thus include transgenic mammals.

Responsibility for mandatory licensing lies with the national authorities or the different Spanish regions, depending on their powers. However, authorization for the placing on the market of products based on genetically modified organisms can only be given if no EU member State has lodged an objection within the stipulated deadline. If no agreement is reached, the European Commission shall give its prior approval (article 20). A licensing procedure followed in another member State suffices for a product to be placed on the market in Sapin (article 21). At the request of the person or company responsible for the activity, certain aspects of the dossier may be treated confidentially (article 23). Lastly, sanctions are established for breaches of the regulations, including the partial or total closure of the premises (article 26 and 27).

Dr. in Laws, Dr. in Medicine. Professor in Criminal Law, University fo the Basque Country. Director, Inter-University Chair BBVA foundation-Provincial Government of Biscay, in Law and the Human Genome, Universities of Deusto and of the Basque Country. Bilbao, Spain. E-mail: cromeo@genomelaw.deusto.es

CHAPTER

3

PLANT HAIRY ROOTS
NATURE'S GIFT TO BIOTECHNOLOGISTS

Susan Eapen & R. Mitra

Agrobacterium rhizogenes is a soil bacterium responsible for the hairy root disease on a range of dicotyledonous plant and a few gymnosperms. The R_1—transfer DNA (R_1—TDNA) which occurs as a single entity or as two regions (TL and TR-DNA) on the bacterial R_1 plasmid is transferred to the plants and gets incorporated into plant DNA. Four regions of TL-DNA are involved in the production of roots and these loci are termed role A, B, C and D (root including locus). Roots that are induced at the site of infection can be excised and cultured *in vitro* and are distinguished from normal roots by their rapid growth, highly branched nature, plagitropic (horizontal) growth and opine synthesis.

Hairy roots are model system in plant metabolic engineering, phytoremediation and as a vehicle for genes for producing transgenic plants.

1. A Source of Secondary Metabolites of Medicinal Value

Plants are a source for the discovery of new products of medicinal value and as lead compounds in the development of drugs. Compounds normally found in hairy roots are those

---

Nuclear Agriculture and Biotechnology Division, Bhabha Atomic Research Centre, Trombay, Mumbai-400 085, India.

synthesized in the roots of intact plants. Hairy roots of *Artemisia* producing antimalarial compound artemisinin, *Catharanthus* producing anti-hypertensive compounds-indole alkaloids, *Glycyrrhiza* producing anti-microbial compounds such as isoprenylated flavanoids and polysaccharides, *Linum* and *Podophyllum* producing anticancer compounds lignans, *Trichosanthes* producing antiviral and antifungal compound—ribosome inactivating protein, *Trigonella* producing steroid diosgenin, *Valeriana* producing tranquilizer valeopotriates, *Digitalis* producing cardioactive glycosides, *Panax* producing saponins, *Withania* producing withanolides, *Azadirachta* producing azadirachtin, *Lawsonia* producing the colouring agent lawsone, and *Lithospermum* producing the red dye shikkonin are a few worth mentioning. It is worthwhile attempting hairy root induction for production of anticancer compounds such as taxol from *Taxus* species and campothecin from *Camptotheca* and *Nothopodytes,* since these products at present are at great demand.

The morphology of transformed root can be affected by different factors such as plant species and *A. rhizhogenes* strains used for hairy root induction location and number of copies of T-DNA integration. There is tremendous variation in the morphological characteristics of roots which is likely to influence the ability to scale up root cultures for commercial production. The relationship between root thickness and nutrient demand can be of fundamental importance to the growth of plant roots in the reactor. Since the root growth differs from microbial growth in several important aspects, there is a need to design special bioreactors for hairy root cultures. Impellers in stirred tanks damage the fragile interconnected root material leading to callus formation Wire meshes can be used to separate roots from impellers, but biomass production is inhibited compared with unstirred vessles. Air-sparged reactors can be used, but has problems too. Alternatively one can move away from submerged culture and design reactors based on strategies such as film flow and nutrient mist.

Elicitation can be used for enhancing bioactive compounds. In addition, biotransformation can lead to development of new compounds. Besides, genes controlling specific pathways leading to the secondary metabolite can be introduced into the hairy root. Expression of genes in an antisense orientation to downregulate a particular enzymic step is also possible. It is possible to increase the availability of substrate for a secondary metabolite pathway by increasing the levels of enzyme at the interphase between primary and secondary metabolism. Like many other advances in technology, successful commercial application will be possible by a co-ordinated multi-disciplinary approach integrating the skills of biochemists, molecular biologists and biochemical engineers.

2. A Source of Therapeutic Proteins and Monoclonal Antibodies

The production of therapeutic proteins by plants is an area of great commercial interest. It is possible to produce many foreign proteins in hairy root cultures by transferring the appropriate genes needed for production. Monoclonal antibodies are currently produced in large scale fermentation systems using hybridoma cells formed by fusion of animal lymphocytes with myeloma cells. However in recent years, several heterologous systems including bacteria, yeast, insect cells and non-lymphoid mammalian cells have been used for antibody production. Plan systems are being considered as an alternative to animal cell cultures as a commercial source of monoclonal antibodies and hairy roots are a suitable candidate. Hairy roots are capable of high levels of production of antibody. It is preferred that antibody is secreted into the medium for better downstream processing. However, problems such as antibody degradation may have to be taken into consideration before large scale production is attempted.

3. Phytoremediation

Environmental contamination with heavy metals comes from enhanced industrial, agricultural and military activities. Green plants can be used to remove and accumulate heavy metals and is an emerging technology promising effective and inexpensive clean up of hazardous waste sites and contaminated areas. Potential application of phytoremediation include bioremediation of petrochemical spills, ammunition wastes, radioactive sites, contaminated chemical storage areas, landfill leachates etc. Hairy roots from hyperaccumulators can be used for studying and understanding the phenomenon of metal accumulation and also probably passage of contaminated solutions through bioreactors growing hairy roots may render the solution with reduced contaminants. Besides, the regenerated plants from hairy roots of hyperaccumulators, because of their altered root architecture may have capacity for higher uptake of heavy metals. It will also be possible to introduce genes from other organisms into plants for better uptake and storage of contaminants.

4. For Modifying Plant Architecture

The altered phenotype induced by *Agrobacterium rhizogenes* is usually regarded as undersirable; but the altered phenotype has potential applications in plant improvement. For eg., for phytoremediation it is essential to have plants with extensive root system. Another major use will be in horticultural industry for altering traits like induction of dwarfing, more branching with more number of flowers and altered leaf morphology. It is also possible to change the photoperiodism and also life cycle-changing perennials to annuals and vice versa, delay or eliminate capacity to flower, which has great implications in some crops. Specific 'rol' genes may result in male sterile lines. In monocots such as grasses and cereals, the specific rol genes can be introduced by particle gun bombardment which will result in increased tillering and dwarf stature. Besides, *A. rhizogenes* can be used for increasing rooting of cuttings and can be applied alongwith conventional rooting mixtures.,

5. For Introduction of Foreign Genes

Foreign genes of agricultural importance can be transferred to plants using *Agrobacterium rhizogenes* as a vehicle. Binary vectors with the desired genes in between T-DNA border elements can be introduced into *A. rhizogenes.* Co-transformation will result in introduction of foreign genes and R_1-TDNA. After regeneration, due to independent insertion of R_1-TDNA and binary vector DNA, it is possible to eliminate the unwanted R_1-TDNA in the next generation

(T_2) by segregation at meiosis of T_1 plants, which will result in normal T_2 plants with the foreign gene. It is easier and faster to get transformation with *A. rhizogenes* in comparison with *A. tumefaciens*.

6. As a Host for Obligate Parasites

The rhizosphere contains obligate parasites that co-evolved with land plants such that they depend on the root for exudation of substrates for growth. Some of these organisms can be cultured using transformed roots as a culture medium. Of the obligate parasites, vesicular-arbucular mycorrhizae (VAM) engage in mutually beneficial association with land plants and contribute to agricultural productivity by mobilizing and translocating phosphorous and other minerals to plants. Hairy roots in bioreactors can be used for large scale cultivation of VAM fungi for distribution to farmers.

Phytopathogenic nematodes can also be grown on hairy roots.

7. In Rhizosphere Research

The rhizosphere is the site of convergence of factors which are critical for the survival of the plant. Environment of the root is conditioned by the physical characteristics of the soil both by assimilation and liberation of substances by the root and by the presence of organisms such as fungi and bacteria. Transformed roots provide a source of axenic material for modelling the rhizosphere phenomena. It would be useful to target a soil bacteria for a plant by incorporating into the plant root selective advantage for growth in the rhizosphere of that species. The study of hairy root will make it possible to understand more about the interactions of roots with rhizosphere microorganisms and plant-microbve interactions.

WORK CARRIED OUT IN BARC

(a) Hairy roots for phytoremediation

Hairy root cultures were initiated from hyperaccumulators such as *Brassica juncea, Chenopodium amaranticolor* and

Helianthus annus. Studies of hairy roots of *B. juncea* with various concentrations of uranium showed that it was capable of taking up uranium from a medium containing 200 uM uranium nitrate without showing any toxic effect to the root tissue (about 7g fresh weight). Uranium was completely depleted from the medium of 50 ml within 10 days of culture. Besides uranium, hairy roots are also being used to remediate heavy metals like Pb and Cd and others like As by the process of rhizofiltration.

(b) Hairy roots for secondary metabolites

Hairy roots were induced from *Azadirachta indica* and *Catharanthus roseus* and secondary products are being investigated.

(c) Hairy roots of leguminous crops

To see if leguminous crops are susceptible to *A. rhizogenes* and subsequently capable of regeneration into plants, hairy roots were induced from important crops such as *Arachis hypogaea, Glycine max, Vigna radiata, Vigna mungo, Vigna aconitifolia, Vigna unguiculata* and *Cicer arietinum*. Attempts are being made to induce differentiation of plants. If differentiation could be achieved, *A. rhizogenes* could be used as a vehicle for gene transfer instead of *A. tumefaciens*.

CONCLUSIONS

Hairy root cultures is "Nature's gift to plant biotechnology". Research on product development from hairy root cultures has progressed from qualitative demonstrations in the past to commercial exploitation. Bioreactors with low capital costs for growing hairy root cultures is the need of the day. With the possibility of introduction of heterologous protein genes into hairy roots, competitions for development of new products from plants has become fierce. The usefulness of heavy metal uptake by hairy roots makes it a starting point to enhance bioaccumulation. *Agrobacterium rhizogenes* also has potential use as a vehicle for gene transfer for development of transgenic plants. The demonstration that 'rol' genes can change plant architecture makes it a unique system for modifying plant traits. The horticulture industry will be one of the first areas to benefit from this Nature's Gift to Mankind.

CHAPTER

4

PLANT BASED PRODUCTION OF VACCINES EDIBLE VACCINES

G.B. Sunil Kumar, T.R. Ganapathi, V.A. Bapat and R.K. Mitra

Introduction

Vaccination is the safest and most economical way to prevent diseases ((Ghosh, 1995) conventional vaccines consisted of live attenuated pathogens, whole inactivated organisms or inactivated toxins. Though they were proved to be successful in the past, several constraints have limited their use against more challenging diseases like Hepatitis B, Cervical cancer and AIDS. Certain live attenuated vaccines can cause disease in immuno-suppressed individuals by reverting to a more virulent form. Whole inactivated vaccines (eg. *Bordetella pertussis*) contain reactogenic components that can cause undesirable side effects. Some pathogens are difficult or even impossible to grow in culture (eg. Hepatitis B, Hepatitis C and Human papilloma virus), making preparation of vaccines problematic (Singh and O'Hagan, 1999).

On a global scenario, the developing countries are often the target of several infectious diseases. Preventive medicine (for eg. Vaccines) has proceeded rapidly in the last decade as

Plant Cell Culture Technology Section, Nuclear Agriculture and Biotechnology Division, Bhabha Atomic Research Centre, Trombay, Mumbai-440 085, India.

biotechnology has been applied in several areas. The requirement of vaccines has been summarised in table 4.1. These medical advances are not likely to have significant impact in developing countries because of prohibitive cost of production and delivery of recombinant proteins, which comprise the new vaccines. The economic conditions and inaccessibility to World markets and limits the use of vaccines for immunization of large population. In this regard, the concept of vaccine production in transgenic plants especially edible vaccines assume greater significance for a relatively low cost, agriculture based system rather than the sophisticated and expensive cell culture and fermentation based vaccine production (Suprasanna *et al.*, 1997).

Table 4.1: Requirement of Important Vaccines**

S. *No.*	*Vaccines*	**Estimated requirement 1994-1995*	**Estimated requirement 1999-2000*
1.	DPT	105	114
2.	DT	50	57
3.	Tetanus toxoid	75	100
4.	BCG	41	43
5.	Polio	105	134
6.	Measles	25	30
7.	MMR	5	7.50
	Rabies:		
8.	Sheep brain based	4	4.5
	Cell Cultured	3	5
	Hepatitis B:		
9.	Plasma derived	0.10	0.20
	Recombinant	1	45
10.	Typhoid (attenuated oral)	10	50
11.	H. Influenza type B	1	5
12.	Meningitis	0.5	2

*Million doses; **Adopted from Ghosh, 1995

Vaccines can be produced in plant parts which can be eaten and therefore called edible vaccines (Mason and Arntzen, 1995; Richter *et al.*, 1996). The concept of producing edible vaccines in plant has been recognised to be attractive for its safety, simplicity and low farming cost (Mason *et al.*, 1992). Vaccines that are effective against infectious diseases must stimulate mucosal immune system as many infectious agents like bacteria and viruses invade epithelial membranes. In general the mucosal immune response is more effectively achieved by oral, rather than parenteral antigen delivery which can be met if the vaccines can be produced in edible tissues of plants. (Meloen *et al.*, 1998).

The successful expression of antigenic proteins in plants has been summarised in Table 4.2. In this report some of the important infectious diseases and their plan based vaccines production has been described briefly.

CASE STUDIES

A. Human Vaccines

1. *Traveller's Diarrhoea*

In the developing countries, diarrhoeal disease is a leading cause of death especially among children. Travellers to these areas are frequently exposed to the bacteria and viruses that cause diarrhoeal disease, resulting in symptoms ranging from discomfort to severe debilitation. Vaccines to prevent diarrhoeal disease could be of value to travellers who lack protective immunity. Endemic bacterial pathogens, including enterotoxigenic *Escherichia coli* (ETEC), which produces a heat labile enterotoxin (LT) is the most common cause of diarrhoea. LT has one A sub unit (LT-A) and a pentamer of B sub units (LT-B). Non-toxic LT-B binds to the G_{M1} gangliosides present on epithelial cell surfaces allows entry of the toxic LT-A sub unit into the epithelial cells of the gut and initiates cellular metabolic changes that lead to loss of water from the cells. LT-B is a potent oral immunogen, when administered orally elicits a strong mucosal immune response without any symptoms of disease. Secretory antibodies in mucosal fluids prevent LT-B binding to the epithelial cells and there by interfere with toxic effect of LT.

Table 4.2: Vaccines produced in Transgenic Plants

Disease Antigen	*Origin*	*Host*	*Production level*	*Reference / Source*
1	**2**	**3**	**4**	**5**
Traveller's Diarrhoea (Heat labile enterotoxin B)	*Enterotoxigenic* E. coli	Tobacco	0.001% soluble leaf protein (S.L.P.)	Haq et al., 1995
		Potao	0.01% S.L.P.	Haq et al., 1995
Hepatitis B (HBsAg)	Hepatitis B virus	Tobacco	0.00666% S.L.P.	Mason et al., 1992
		Potato & Tomato	***	***
Gastroenteritis (Norwalk virus capsid protein)	Norwalk virus	Tobacco Potato	0.23% S.L.P. 0.37% S.L.P.	Mason et al., 1996
Food and Mouth disease (VP1)	Papilloma virus	*** alfalfa	*** ***	Usha et al., 1993 Wigdorovitz et al., 1999

(Table Contd...)

Table 4.2: (Contd...)

1	2	3	4	5
Immunoco-ntraception (Zona Pellucida Protein 3)	Murine	Tobacco	***	Fitchen et al., 1995
Malaria (epitopes derived from sporozoites)	Plasmodium sporozoites	Tobacco	***	Turpen et al., 1995
Mink enteritis (VP2)	Mink enteritis virus	Cowpea	***	Dalsgaard et al., 1997
AIDS (gP 41)	HIV	***	***	Porta et al,. 1994
Swine fever (Hog cholera, E0, E1 & E2)	Swine fever virus	Tobacco	***	In: Kapusta et al., 1999
Rabies (Glycoprotein)	Rabies virus	Spinach	***	Modelska et al., 1997
Swine Transmissible Gastroenteritis	Transmissible Gastroenteritis virus	Corn	***	In: Biotech International 2000

(Table Contd...)

Table 4.2: (Contd...)

1	2	3	4	5
Dental caries (SpA antigen)	*Streptococcus* mutans	Tobacco	0.02% S.L.P.	In: Mason & Arntzne, 1995
Auto immune diabetes (Insulin)	Human Pancreas	Potato	0.05% S.L.P.	Arkawa et al., 1998
Cholera (CT-B)	*Vibrio Cholerae*	Tobacco Potato	***	Arakawal et al., 1998? Hein et al., 1996
Colon cancer	***	Tobacco	***	Verch et al., 1998
Influenza	Influenza virus	Tobacco	***	Beachy et al., 1996
Lymphoma (Tumor derived ScFv epitopes)	***	Tobacco	***	Mc Cormick et al., 1999
Post surgical/ burn infections outer membrane protein F	*Pseudomonas aeruginosa*	Tobacco	***	Stackzek et al., 2000
Common cold (VP1)	Human Rhinovirus	***	***	Porta et al., 1994

Recently it has been shown that same amount of rLT-B expressed in *E. coli* and also in potato when administered to two groups of mice, both groups of mice developed equal amounts of antibodies in the serum and mucosa and the biological activity of LT was neutralised by these antibodies to the same extent (Haq *et al.*, 1995). In the first human clinical trials of edible vaccine, LT-B antigen delivered through transgenic potato resulted in serum and/or mucosal immune response (Tackett *et al.*, 1998). This first trial is a milestone on the road to create inexpensive vaccines that might be particularly useful in immunizing people in developing countries, where high cost and logistical issues and the need for certain vaccines to be refrigerated, can thwart effective vaccination.

2. *Gastroenteritis*

Norwalk virus causes epidemic acute gastroenteritis in humans. Expression of Norwalk capsid protein (NVCP) in insect cells yields a protein with an apparent M of 58,000 that self assembles into insect cell derived Norwalk virus like particles (i-rNVs) lacking viral RNA which are reactive with sera from Norwalk virus infected humans. This led to the expression of recombinant NVCP in transgenic tobacco leaves and potato tubers. The plant expressed rNV was orally immunogenic in mice. Extracts of tobacco leaf expressed rNV when administered to mice developed both lgG and secretory IgA specific for rNV (Mason *et al.*, 1996).

3. *Hepatitis B*

Hepatitis B is one of the major world wide viral diseases and probably the single most important cause of persistent viraemia in humans. Current estimates establish that there are about 300 million chronic carriers of Hepatitis B virus (HBV) which constitute 5% of total world population. 20 million new infections occur annually. The late sequel of HBV infection is hepatocellular carcinoma. About one million new cases of hepatoma and 350,000 deaths from hepatoma occur annually. In India 2,70,000 children are afflicted every year by this virus. While its treatment is costly, the cure rate is not impressive. Antibodies to hepatitis B surface antigen (HBsAg) confers protection against HBV infection. Since HBV can not be cultured in tissue culture and the host range is limited to chimpanzees and humans, the first vaccine consisted of HBsAg purified from the plasma of carriers. However, concerns about the safety resulted in the development of recombinant vaccine, produced by expression of HBsAg in yeast.

In many developing countries, the expense of immunisation programme prohibits the use of currently available vaccine for large segments of population. This led to endeavour the expression of HBsAg in plants with the hope of developing a less expensive product. The first report demonstrating the feasibility of plant based vaccine production was reported by Mason *et al.,* (1992). Further the same group reported that rHBsAg self assembles into sub viral particles which are virtually indistinguishable from the plasma and yeast derived HBsAg with respect to size, density sedimentation, antibody binding and in eliciting HBsAg specific antibodies in mice and prime T cells in vivo that can be stimulated in vitro by the yeast derived rHBsAg.

Currently, our group is engaged in a collaborative research programme with Shantha Biotechnics Pvt. Ltd., Hyderabad, to develop edible vaccines for Hepatitis B by expressing HBsAg in banana and tomato plants. In banana we have already developed the successful protocol for agrobacterium mediated transformation using embryogenic cells (Ganapathi *et al.,* 1999, 2000).

4. *Cervical cancer*

Cervical cancer is the second most common malignant tumor in the World and ranks first in Indian women. In India about 100 thousand women develop this cancer every year constituting about 16% of the world's annual incidence. Human papilloma virus (HPV) is unequivocally proved to be associated with cervical cancer. In India, research has not been conducted on HPV vaccines and hence concerted research efforts in this area are warranted (Das *et al.,* 2000). Mucosal immunity will be of prime importance in conferring protection against cervical cancer. In this direction, edible vaccine strategy would be the most appropriate one since it works as a simple inexpensive and painless delivery system.

Currently, our group is working on a collaborative project with Cancer Research Institute, Mumbai to develop plant based production of vaccines for Cervical cancer.

5. *Cholera*

Cholera is a devastating infectious diarrhoeal diseases throughout the World. The total number of annual cholera cases exceeds 5 million and around 200,000 people die due to this disease. Effective prevention of cholera dissemination depends on safe water sources and improved sanitation which many of the developing countries can not afford. The alternative is to develop oral vaccines, that provide lasting protection after a single immunization or convenient and readily available vaccines for frequent administration to people living in regions where cholera is endemic. Since cholera is controlled more effectively through mucosal rather than parenteral immunization, the edible plant based production of cholera vaccine will be easy for administration. In this direction reports are available on the expression of CTB in potato (Arakawa *et al.,* 1998). Transgenic potatoes were engineered to synthesize CTB pentamer. Further it has been noted that both serum and intestical CTB-specific antibodies were induced in orally immunized mice. The expression of oligomeric CTB protein with immunological and biochemicals properties identical to native CTB protein in edible plants like potato opens the way for preparation of inexpensive food plant-based oral vaccines for protection against cholera (Arrakawa *et al.,* 1997).

B. Veterinary Vaccines

1. *Food and mouth disease of cattle*

(FMD) is an acute, highly contagious *Picornavirus* infection of cloven hooved animals. FMD is present in many parts of the world. The disease may spread over long distances with movement of infected or contaminated animals, products, objects, and people. Pigs are mainly infected by ingesting infected food. Waste feeding has been associated with outbreaks. Cattle are mainly infected by inhalation, often from pigs, which exrete large amounts of virus by respiratory aerosols and are considered highly important in disease spread. Large amounts of virus are excreted by infected animals before clinical signs are evident, and winds may spread the virus over long distances.

Clinical signs in cattle are salivation, depression, anorexia and lameness caused by the presence or painful vesicles (blisters) in the skin of the lips, tongue, gums, nostrils, coronary bands, interdigital spaces and teats. Fever and decreased milk production usually precede the appearance of vesicles. The vesicles rupture, leaving large denuded areas which may become secondarily infected. In pigs, sheep and goats the clinical signs are similar but milder. Lameness is the predominant sign.

FMD is one of the most alarming disease, because of the range of spices affected, the high rate of infectivity, and the fact that virus is shed before clinical signs occur. An outbreak of FMD would, (and has in the past) cost millions of dollars in lost production, loss of export markets, and loss of animals during eradication of the disease (in *"Food-and-Mouth Disease Strategy"* Agriculture and Agri-Food Canada, November 1994). An effective vaccine will help in reduction of disease incidence.

Mice parenterally immunised using leaf extracts of transgenic alfalfa plants expressing structural protein VP1 of food and mouth disease virus developed a virus specific immune response. The immunised mice were protected when challenged with the virus (Wigdorovitz, *et al.,* 1999).

2. *Swine transmissible Gastroenteritis*

Caused by Swine transmissible gastroenteritis virus (TGEV). TGEV can infect pigs of all ages, but mortality is high in pigs which are under two weeks old. Such early infection can result in substantial losses to the pork industry. TGEV is a well-characterized model of mucosal immunity in swine. The S protein of TGEV forms a large termitic structure protruding from the viral envelope and likely facilitates the attachment to the epithelial cells lining the intestine. It is also the major antigen responsible for the generation of neutralizing IgA antibodies, which are secreted from the epithelia, thereby inhibiting viral attachment. Antibodies can also be secreted in the milk of a sow which has been infected or inoculated, thereby providing passive immunity to suckling piglets. Current preventative therapy involves orally vaccinating sows with live virus, thereby triggering an immune response that generates neutralizing antibodies in milk and colostrum. These antibodies provide passive immunity to suckling piglets. There is an increasing interest in developing a plant-based TGEV vaccine delivery system. Prodigene, a USA based company reported that clinical trails with an edible vaccine expressed in corn can give protection against virulent TGEV (In Biotech International, 2000).

STRATEGIES FOR VACCINE PRODUCTION IN PLANTS

(a) Stable expression by developing transgenic plants

In this strategy, antigen coding gene from pathogen will be cloned into plant expression vector with appropriate promoter for systemic/tissue specific (fruit/seed specific) expression. These vectors will be used to transform plants through *Agrobacterium* mediated or biolistic transformation or any other method used for stable transformation.

(b) Transient expression by using plant viral vectors

In this type, antigen/epitope coding gene can be expressed as a chimeric gene for viral coat protein using viral vectors or foreign gene transcription driven from sub genomic promoter. Currently Tobacco mosaic virus (TMV), Cow pea mosaic virus (CPMV) and Tomato bushy stunt virus (TBSV) based vectors are in use. *In vitro* transcribed RNA of the above chimeric viruses is used to inoculate plants. This strategy yields particulate vaccines, which are highly immunogenic and confers protective immunity (Koo *et al.*, 199, Stackzek *et al.*, 2000). Partially purified recombinant virus particles can be used for oral administration.

The following factors play a pivotal role in successful development of pland based vaccines:

1. Choice of Crops for Developing Edible Vaccines

a. *Human vaccines*

For human Vaccines, it is appropriate to use fruit and vegetable crops, which can be eaten raw. Amongst these bananas offer multiple advantages including their availability in tropical and subtropical countries in the developing world, where many vaccines are needed. They are richly available at lower cost and consumed as staple food and suitable in terms of digestibility and palatability by infants. Majority of dessert bananas are triploid and are propagated vegetatively. Therefore it is an ideal crop for biological containment. It is estimated that approximately 20,000 tonnes of bananas are needed to vaccinate all the world's children which is only a fraction of the total world production. One banana can potentially contain ten doses of vaccine (Arntzen, 1997). Delivery of vaccines in bananas may enhance their immunogenicity, as their pulp contains lectins which serve as mitogens for lymphocytes (Pneumans *et al.*, 2000).

(b) *Veterinary vaccines*

Legumes, cereals and forage crops would be the suitable candidates, as they are used as animal feed. Utilisation of Seeds have some advantages like they have low level of endogenous proteases and protective coat as a result immunogenic proteins can be stored for a long time without any cold storage. They are referred as biological containers.

2. Mucosal Adjuvants

Edible vaccines delivered through oral route are exposed to acid in the stomach, which many limit the amount of orally delivered vaccine that reaches the gut associated lymphoid tissue. Vaccine adjuvant therefore plays important role to enhance the immunogenicity of orally delivered antigens. Bacterial antigens like enterotoxigenic *E. colil* heat labile toxin (LT) and cholera toxin (CT) are strong mucosal adjuvants. Mutant forms of these toxins like LTK-63 etc., retain their adjuvanticity with lowered toxicity (Rappuoli *et. al.*, 1999). These mutant toxins can be employed as adjuvants for edible Vaccines.

3. Optimization of Gene Expression to increase the level of Antigen Expression in Plants

(a) Targeting of recombinant problems

The recombinant proteins when targeted either to endoplasmic reticulum or secretory pathway showed proper folding of proteins, thereby increasing functional protein level expressed in plants (Pen *et al.*, 1993a; Schouten *et al.*, 1996). For example ER retention signal can give 10-100 fold increases in target protein yield compared to secretory pathway (Conrad *et al.*, 1998). ER targeting is essential for glycosylation and disulfide bridge formation (Lturiga *et al.*, 1989; Bruyns *et al.*, 1996). Transgenic plants expressing ER retained single chain variable fragments (ScFvs) have high antign binding activity and can be stored for more than 3 weeks without loss of antign binding activity or specificity in a dried form (Fiedler *et al.*, 1997). Mason and Arntzen (1995) reported that recombinant LT-B produced in tobacco and potato showed enhanced accumulation when a C terminal microsomal retention signal was added. McCormick used rice alpha amylase signal peptide to fuse ScFv sequence of B-cell lymphoma epitope targeted the protein into the secretory pathway.

(b) Modifying the target gene according to the host plant preferred codons

By using plant preferring codons one can increase the level of target protein expression. The *Bacillus thuringiensis* crystal protein genes were found to be poorly expressed in plants (Fischoff, 1992). Failure of the expression appear to be due to the coding sequence itself. The GC content of bacterial genes is generally less than that of plants and they may contain AT sequences which serve as polyadynylation signals of eucaryotic genes. Construction of synthetic genes with preferred codon usage of the host plant will enhance the expression levels (Sutton *et al.*, 1992). Similarly synthetic gene with a plant preferring codons, encoding LT-B has been used to increase the LT-B expression level in transgenic plants.

4. Post Translational Modifications in Plants

Among the post translational modification, glycosylation has been shown to play critical roles for various physiological activities of mammalian glycoproteins (Kusnadi *et al.,* 1997; Pen *et al.,* 1996). Properties of recombinant proteins such as biological activity, protein folding, stability and solubility could be affected by glycosylation (Marino, 1991). The high mannose type N-glycans in plants have structures identical to those found in other eucaryotic cells, however, plant complex N-linked glycans differ substantially (Chrispeels *et al.,* 1996). In the mammalian system, The Man_3 (M3) core structure is extended further to contain penultimate galactose and terminal sialicacid residues (Kornfeld *et al.,* 1985). In contrast, typically processed N-glycans in plants are mostly of Man_3 $GlcAc_2$ structure with or without B 1, 2 xylose and or × 1, 3 fucose residuces (Chrispeels *et al.,* 1996; Rayon *et al.,* 1998). Larger complex type plant N-glycans are rare and recently have been identified as additional × 1, 4 fucose and B 1,3 galactose residuces giving rise to mammalian lewis a structure (Lerouge *et al.,* 1998). The presence of xylose and/or fucose residues makes plant recombinant therapeutics less desirable as they induce immune responses in mammals. The transfer of human B 1, 4-galactosyltransferase to tobacco modified the N-linked glycosylation pattern resulting in glycans with galactose residues at the terminal non-reducing ends. In addition, the absence of the dominant xylosidated and fucosylated type sugar chains confirms that the transformed cells can be used to produce glycoproteins without the highly immunogenic glycans (Palacpac *et al.,* 1999).

CONCLUDING REMARKS

Currently majority of the vaccines are being administered using needle or syringe subcutaneously or intramuscularly. Only very few like plio, typhoid and cholera Vaccines are given orally. Future generation of vaccines should focus more heavily on mucosal immunisation. Edible vaccines activate the mucosal immune system against many pathogens by oral or nasal delivery. The use of injections can be avoided, therefore improving patient compliance, reduces the risk of transmission of infection through contaminated needles and administration costs. They contain antigens rather than whole pathogens, thus eliminating the risk of causing the disease, the vaccine is intended. They may not require cold storage, therefore increasing shelf life. Can be produced cost effectively and in large quantities in agriculture based farming. Since plants do not contain human pathogens and plant viruses are not shown to infect humans, plant system serves as an ideal alternative to the existing microbial/ animal cell cultures in terms of safety. Beyond these pragmatic advantages is the desirability to exploit the mucosal immune system. Mimicry of the natural route of infection (oral, respiratory and genital) assures the exposure of large surface areas to vaccines, the development of a first line defence mediated by secretory immunoglobulin a, recruitment of a humoral antibody response via lymphocyte trafficking from mucosal lymphoid tissues and involvement of glandular epithelium linked to the mucosal system.

However a number of considerations to be taken into account in the development of any recombinant vaccines in plants. These include fidelity of the antigen, in terms of "antigenicity and immunogenicity which in turn, depends on folding, structure and glycosylation; antigen stability; ease of purification, the potential for scale up production to produce sufficient quantities, cost and safety issues. Further maximising the expression of the antigenic proteins, stabilising the foreign protein during post harvest storage in plant tissue and enhancing the oral immunogenicity of some antigens have to be determined and when administered orally, it has to withstand harsh conditions, high dilution and needs to be transported in intact form across the epithelial barrier in sufficient amounts. Passage of large molecules across epithelial barrier seems to be possible under special circumstances like malnutrition, disease or by some components in food like saponins. Concerns about allergy and oral tolerance have to be addressed.

REFERENCES

1. Arakawa T., Chong, D.K.X., Merrittjl and Langridge, W.H.R., 1997, "Transgenic Res.' 6: 403-413.

2. Arakawa T., Yu J., Chong, D.K.X., Hough J., Engen P.C. and Langridge W.H.R., 1998, Nat. Biotech. 16:934-938.

3. Arntzen C.J., 1997, In: infectious diseases in children speciality forum, Slack Incorporated.

4. Beachy R.N., Fitchen J.H. and Hein M.B. (1996) Ann. New York Acad. Sci. 792: 43-49.

5. Bruyns, A.M., de Jaeger, G., de Neve, M., de Wilde, C., van Montagu, M. and Depicker, A. (1996) FEBS Lett. 386: 5-10.

6. Chirspeels, M.J. and Faye, L (1996) In transgenic plants: A production system for Industrial and Pharmaceutical proteins, eds. Owen, M.R.L. and Pen, J (Wiley, New York), 99-113.

7. Conrad, U and fiedler, U. (1998) Plant Mol. Biol. 38: 101-109.

8. Crammer C.L., Boothe J.G. and Oishi K.K. (1999) Transgenic plants for therapeutic proteins: Linking upstream and down stream strategies In: Current topics in Microbiology and Immunology, Vol. 240 Plant Biotechnology: New products and applications. Hammond J., McGarbvey P. and Yusibov V. (Eds.). Springer—Berlag, Berlin Heidelberg. pp. 95-118.

9. Daalsgard, K. Uttenthal, A., Jones, T.D., Xu, F., Merryweather, A., Hamilton, W.D.O., Langeveld, J.P.M., Boshuizen, R.S. Kamstrup, S. Lomonosoff, G.P. Porta, C., Vela, C., Casal, J.I., Meloen, R.H. and Rodgers, P.B. (1997), Nature Biotechnology. 15: 248-252.

10. Das, B.C., Gopalkrishna, V., Hedau, S. and Katiyar, S. (2000) Curr. Sci. 78(1): 52-62.

11. Draper J., Scott, R. and Hamil J. (1988) Transformation of dicotyledonous plant cells using the Ti plasmid of Agrobacterium tumefacies and Ri plasmid of A. rhizogenes. In: Plant genetic transformation and gene expression A laboratory manual. Draper, J., Scott R., Armitage P. and Walden R. (Eds). Blackwell Scientific Publications. pp. 69-160.

12. Fiedler, U., Phillips, J., Artsaenko, O. and Conrad, U. (1997) Immunotechnology, 3: 205-216.

13. Fischoff, D.A. (1992) J. Cell Biochem: Suppl. 16F: 199.

14. Fitchen, J., Beachy, R.N. and Hein, M.B. (1995) Vaccine 13: 1051-1057.

15. Ganapathi T.R., Higgs N, Van Eck J, Balint-Kurti PJ and May GD. (1999), Transformation and regeneration of the banana cultivar, Rasthali (AAB). In: International symposium on Molecular and Cellular biology of banana. March 22-25, 1999, Ithaca, NY, USA. pp. 34.

16. Ganapathi TR, Higgs NS, Balint-Kurti PJ, Arntzen CJ, May GD and Van Eck JM (2000), *Agrobacterium*-mediated transformation of the banana cultivar Rasthali (AAB). Plant Cell Reports (In Press).

17. Ghosh P.K (1995) Role of biotechnology in health care in India: Present and Future. In: Biotechnology and development Review. Biotech Consortium India Ltd. pp. 1-22.

18. Hansen E. (1997), Brazilian J. Genet 20:703-711.

19. Haq T.A., Mason H.S., Clements J.D. and Arntzen C.J. (1995) Science 268: 714-716.

20. Ituriaga, G., Jefferson, R.A. and Bevan, M. (1989) Plant Cell 1:381-390.

21. Kapusta, J., Modelska, A., Figlwrowicz, M., Podkowinski, J., Priewski, T., Losowa, O., Berbezy, P., Helias, D., Biesiadka, J., Femiak, I., Letellier, M. Yusibov, V., Plucienniczak A., Koprowski, H. and Legocki, A.B. (1999) A. Altman et al. (eds.) Plant Biotechnology and Invitrobiology in the 21st century. pp. 571-574. Kluwer Academic Publishers, Netherlands.

22. Koo, M., Bendahmane, M., Lettieri, G.A., Paoletti, A.D., Lane, T.E., Fitchen, J.H. Buchmeier, M.J. and Beachy, R.N. (1999) Proc. Nat. Acad Sci. (USA), 96: 7774-7779.

23. Kusnadi, A.R., Niklov, Z.L. and Howard, J.A. (1997) Biotechnol Bioeng 56:473-484.

24. Kornfeld, R and Kornfeld, S. (1995) Annu. Rev. Biochem. 54: 631-664.

25. Leclerc C. and Ronco J. (1998) Immunology Today 19:300-302.

26. Lerouge, P., Cabanes-Macheteau, M., Rayon, C, Fischette Laine, A.C., Gomord, V. and Faye L. (1998) Plant. Mol. Biol. 38:31-48.

27. Lyons P.C., May, G.D. Mason H.S. and Arntzen C.J. (1996) Pharma, News 3: 7-12.

28. Marino, M.W. (1991) In: A. Prokop, R.K. Bajpal and C.S. Ha (eds). Recombinant DNA technology and applications, pp. 29-65. Mc. Graw-Hill, New York.

29. Mason, H.S., Lam, D.M.K. and Arntzen, C.J. (1992) Proc. Nat, Acad. Sci. (USA) 89: 11745-11749.

30. Mason H.S. and Arntzen C.J. (1995) Trends in Biotech. 13: 388-392.

31. Mason H.S., Ball, J.M., Shi J.J., Jiang X., Estes M.K. and Arntzen C.J. 1996) Proc. Nat. Acad. Sci. (USA) 93: 5335-5340.

32. Mc Cormick, A.A., Kumagai, M.H., Hanley, K., Turpen, T.H., Kakim I., Grill, L.K., Tuse, D., Levy, S. and Levy, R. (1999) Proc. Nat. Acad. Sci. (USA) 96: 703-708.

33. Meloen R.H., Hamilton W.D.O., Casal J.I. Dalsagaard K. and Langeveld J.P.M. (1998). The Veterinary Quarterly 220: S92-S95.

34. Modelska, A. et. al., (1997) Proc. Nat. Acad Sci. (USA) 95: 2481-2485.

35. Palacpac, N.Q., Yoshida, S., Sakai, H., Kimura Y., Fujiyama, K., Yoshida, T. and Seki, T. (1999), Proc. Nat. Acad. Sci. (USA) 96: 4692-4697.

36. Pen, J., Sijmons, Van Ooijen, A.J.J. and Hoekema, A. (1993) In: A. Hiatt (ed.) Transgenic plants fundamentals and applications. Marcel Dekker, New York. pp. 239-251.

37. Pen, J. (1996) In transgenic plants: A production system for Industrial and Pharmaceutical proteins, eds. Owen, M.R.L. and Pen, J. (Wiley, New York) 149-167.

38. Pneumans, W.J., Zhang, W., Barre, A., Astour, C.H., Kurti. P.J.B.K. Rovira, P. Rouge, P., May, G.D., Leuven, F.V., Bachi, P.T. and Damme J.M.V. (2000), Planta 211: 546-554.

39. Porta, C., Spall, V.E., Loveland, J., Johnson, J.E., Barker, P.J. and Lonomossoff, G.P. (1994) Virology 202: 949-955.

40. Rappuoli, R., Pizza, M., Douce, G. and Dougan, G. (1999) Immunol. Today 20: 493-499.

41. Rayon, C., Lerouge P. and Faye, L. (1998) J. Exp. Bot. 49, 1463-1472.

42. Richter L., Mason H.S. and Arntzen C.J. (1996) J. Travel Med. 3: 52-56.

43. Schouten, A., Roosien, J., Van Engelen, F.A., de Jong, G.A.M., Borst, Vrenssen, A.W.M., Zilverentant, J.F., Bosch, D., Stiekema, W.J., Gommers, F.J., Schots, A. and Bakker, J. (1996) Plant Mol. Biol. 30: 781-793.

44. Singh M. and O'Haggan D. (1999) Nat. Biotech. 17:1075-1081.

45. Stackzek, J., Bendahmane, M., Gilleland, L.B., Beachy, R.N. and Gilleland Jr., H.E. (2000) Vaccine 18: 2266-2274.

46. Suprasanna, P., Ganapathi, T.R. and Rao, P.S. (1997). Curr. Sci. 72(1): 7-9. (Research News).

47. Sutton, D.W., Havastad, P.M. and Kemp. J.D. (1992) Transgenic Research 1: 228-236.

48. Tacket, C.O., Mason, H.S., Losonky, G., Clements, J.D., Levine, M.M. and Arntzen C.J. (1998) Nature Medicine 4: 607-609.

49. Thanavala Y., Yang Y.F., Lyons P., Manson H.S. and Arntzen C.J. (1995) Proc. Natl. Acad. Sci. (USA) 92: 3358-3361.

50. Turpen, T.H., Reml., S., Charoenvit Y., Hoffman S.L., Fallarme V. & Gr. ill L.K. (1995) Biotechnology 13:53-57.

51. Usha R., Rohll J.B., Shanks M., Maule A.J., Johnson J.E. and Lomonosoff G.P. (1993) Virology 197: 366-374.

52. Verch, T., Yusibov, V. and Koprowski, H. (1998) J. Immunol. Methods 220: 69-75.

53. Wigdoroviz, A., Carrillo, C., Dus Santoz, M.J., Trono K., Peralta, A., Gomez, M.C., Rios, R.D., Franzone, P.M., Sadir, A.M., Escribano, J.M. and Borca, M.V. (1999) Virology 255(2): 347-353.

54. Edible Corn based vaccine for live stock virus (2000). In Biotech international 3-4: 3.

CHAPTER

5

MICROORGANISMS AND GLOBAL SUSTAINABILITY OF PLANT INDUSTRY

P.N. Shastri

Microogranisms—the microscopic life forms—as a group, represent extremely diverse metabolic properties in spite of their primitive morphology, and are an important constituent of ecosystem. Their diverse activities have been observed and utilized by the mankind since prehistoric times. Systematic investigations in 18th and 19th century led to better understanding of their positive and negative roles, specially in the field of medicine and food processing. Developments in biochemical engineering paved the way for the commercial exploitation of their metabolic potential, leading to the development of biotechnology as an alternative ecofriendly and sustainable technology for the new millenium. More and more industries are trying to identify the exact role played by microorganisms, finding the ways to control their undesirable activities and assessing the significance of desirable activities. Paint industry, in recent years, has started to realize the importance of microorganisms for global sustainability, and is trying to explore their constructive and destructive activities, as well as the potential of biotechnology in this area. The positive and negative roles of microorganisms in paint industry are summarised in Table 5.1.

Head, Department of Food Technology, Laminarayan Institute of Technology, Nagpur (M.S.), India.

Table 5.1: Desirable and undesirable activities of microorganisms in relation to paint industry

Desirable activities	*Undesirable activities*
Detoxification	Biodeterioration
Decontamination	Paint in can
Bioremediation	Paint film
Biodegradation of VOC	Corrosion
Formation of paint additives	Biocide Resistant organisms
Antifouling agent	Acquired resistance to biocides
Gelling gent	
Pigments	
Surfactants	

Microorganisms as biofouling agents

The microbial activities involve undesirable biodeterioration/biofouling, which is defined as undesirable changes brought about by the living organisms in the quality or value of the material, either in aesthetic or utilitarian terms. They are involved in spoilage of paints in cans during storage, as well as in the fouling of paint films after the application (Table 5.2 and 5.3). The course of fouling and the microorganisms involved in fouling varies according to the

Table 5.2: Types of biofouling in paints

A. Effect of microbial attack on paint in the can

Ingredient affected	*After effects*
Cellulosic thickner	Viscosity drop, gas production, pH shift
Surfactants	Poor hiding, colour shifts, precipitation of pigment, gelling
Coalescing agents	Gloss reduction, poor fridge thaw stability, chalking, porous film, poor adhesion, poor flow and levelling
Defoamers	Foaming, porous film
Dispersed colour	Off colour, agglomeration, uneven colour

Table 5.3: Effect of microbial attack on paint film

Location	*Agent*	*Deterioration*
Interior	Fungi Yeast Bacteria	Discolouration (grey, green, black) slimy growth. Growth and penetration of hyphae causes cracking, loss of adhesion, blistering, production of metabolites
Exterior	Algae/lichens	Algae growth causes discolouration (green to black) humidity retention leading to biofilm formation, Secondary invasion, cracking by freeze thaw

location and the environment in immediate contact with the paint films. (Table 5.4 and 5.5). The primary fouling agents on aquatic surfaces exposed to light happen to be algae and diatoms, where as fungi and bacteria follow the trend forming a biofilm on the surface. This film further attract the

Table 5.4: Microorganisms commonly encountered with paint defects

Bacteria	*Fungi*	*Algae*
Achromobacter sp	Aspergillus sp.	Blue algae
Aeromonas sp.	Aureobasidium sp.	Oscillotoria
Alcaligenus sp.	Cladosporium sp.	Nostoc
Bacillus sp.	Penicillium sp.	Phormidium-fovedanum
Eschericha sp.	Streptomyces sp.	Scytonema
Proteus sp.		Green algae
Proteus sp.		Chlorella vulgaris
Pseudomonas sp.		Chlorella pyrenoidosa
		Scenedesmus sp.
		Brown algae
		Tribonema aequale

Table 5.5: Course of Biofouling at various locations

(i) Course of biofouling of exterior paint film

Availability of sunlight and CO_2

↓

Growth of algae and diatoms

↓

Formation of plaque

↓

Availability of organic nutrients

↓

Growth of other heterotropic bacteria

↓

Formation of biofilm containing microbial consortium

↓	↓
Aquatic environment Attachment of molluscs, Growth of barnacles	Terrestrial environment Blistering, cracks, corrosion

(ii) Course of biofouling of interior paint film

Availability of organic matter and O_2

↓

Growth of fungus/yeasts

↓

Formation of plaque

↓

Growth of other heterotropic bacteria

↓

Formation of biofilm containing microbial consortium

↓

Development of cracks, blistering

crustaceans and molluscs, which are responsible for macrofouling. The components of biofilm do not penetrate the paint layer, but cause corrosion due to their metabolic end products which are acidic. This leads to cracking/blistering of the surface, facilitating the chemical corrosion. The primary fouling agent on internal paints surfaces are fungi, which then attract the insects for further attack. The course of fouling and the microorganisms involved in fouling varies according to the location and the environment in immediate contact with the paint films.

The paint biocide is subsequently defined as the agent that can control the biodeterioration of paints, when incorporated in it. An ideal biocide should prevent the fouling efficiently, without any harmful effect on the ecosystem, and still be cost effective. (Table 5.6) Various 'biocides' are added to the paints to prevent the deterioration of paints films (Table 5.7). Concern is being expressed about the organo tin

Table 5.6: Desirable properties of paint biocides

1. Compatible with paint formulation.
2. Stable to light, temperature and pH.
3. Board spectrum activity.
4. Effective in low concentration.
5. Should not give discolouration to paint film.
6. Biodegradable, when discharged in an effluent stream and not biomagnified in the ecosystem.
7. Non toxic to users.
8. No leaching from dried/cured films.
9. Adequate migration rate for effective biocidal activity.
10. Slow leaching rate for residual activity, and long time span to reach mcc from mac
11. Last but not the least, cost effective.

Table 5.7: Site of action of some paint biocides

S.No.	*Chemical nature of Biocide*	*Observed site of action*
1.	Phenols and phenolcs	Cytoplasmic membranes, Cytoplasmic proteins
2.	Alcohols	Cytoplasm.
3.	Halogens	Cytoplasm, Cytoplasmic membranes, Proteins
4.	Heavy Metals	Cytoplasm, Cytoplasmic membranes, Nuclear material
5.	Soaps and Detergents	Cytoplasmic membranes
6.	Quart, Metal Compounds (Tributyl tin)	Cytoplasmic membranes, Cell Wall
7.	Aldehydes	Cytoplasm, Cytoplasmic membranes, Cell Wall.

compounds presently being used as antifouling agents in the aquatic paints. An ideal biocide which is non existent is expected to have selective toxicity against fouling microbes, low release rate, 'biodegraded to non toxic components, compatible with the paints components and cost effective. Mercury and tin based biocides are shown to accumulate in the marine environment. As it passes through the food chain, the ultimate effects can be disastrous. Several patents exist for metabolic, heterocyclic and/or polymeric synthetic compounds, effective against algae.

Microorganisms for pollution control

Volatile organic compounds (VOC) as well as the solvents in the paints are identified as the pollutants, that need to be eliminated from the effluent stream. Microorganisms are being examined as potential agents to participate in the effluent treatment in various ways, such as:

- Decontamination of air containing VOC,
- Bioremediation of the hazardous compounds present in the effluent from paint industry.
- Treatment of spray paint sludge and its conversion to form accessible for subsequent Soil remediation.

Volatile organic compounds (VOC) emission control

Volatile organic compounds such as CH_3OH, CH_2Cl, C_6H_6, $CHCl_3$, $CH_3COOC_2H_5$, methyl cellosolve etc. are emitted in the environment. Chemical conversion methods such as thermal and catalytic conversion, incineration, ozonisation and chlorination have the disadvantage of high energy requirement and use of chemicals. Physical processes such as adsorption/absorption do not convert the pollutants, and generate a new polluted mass during regeneration process. Biofiltration process depends on the ability of microorganisms to degrade wide variety of organic compounds for energy generation. Several VOCs can be eliminated from the discharged air with the help of a biofiltration column containing ligocellulosic matter which provides the anchor for the microflora as well as holds moisture and nutrients.

Microflora consists of *pseudomonas putida* along with the composting organisms, basic process is represented as:

$$\text{Pollution} \xrightarrow{\text{microbial enzymes}} CO_2 + H_2O + \text{energy}$$

An experimental column used for the elimination of methylene chloride from the air indicated that with acclimatisation, the microflora could degrade almost 98% pollutant from stream containing 40 ppm of dcm., the process has the advantage of being ecofriendly, and cost effective. It is indicated that 60 out of 189 VOCs could be treated in the biosystems.

Bioremediation

Bioremediation is a technique of adding nutritive material to contaminated environment to accelerate the natural biodegradation process. It comprises two parts.

1. Fertilisation : addition of nutrients like N & P source,
2. Seeding : addition of specific microorganisms

A bioremediation effort at Asian Paints Ltd. reports the isolation of 26 cultures from the preliminary screening from spray booth water and composite soil sample, capable of degrading mineral oils and volatile aromatics present in wet paint sludge and hazardous solvent. Five cultures were finally selected after the enrichment and adaptation to higher concentrations of about fifteen solvents in kazuyoshi medium, and elimination of slow growers. Artificial sludge containing equimix of 6 coatings, 30-50% VOCs and binders, when treated with the consortium, gave the reduction in COD, odour and VOC. Aerobic or anaerobic treatments are being investigated for their utilisation in effluent treatment in paint industry. Some of the microorganisms found to be useful in bioremediation are listed in (Table 5.8). Thus a synergistic combination of microbial cultures isolated as a consortia can be successfully employed in:

(a) Treatment of solvent laden/bearing liquid effluent for COD reduction.

(b) Treatment of the unmanageable spray paint sludge and convert it into a form accessible for subsequent soil remediation.

(c) Effective reduction of VOC in the effluents gas streams exhausts of spray booths.

Table 5.8: Micro organisms useful in bioremediation

Compounds	*Micro-organisms*
Aliphatic hydrocarbons	Pseudomonas sp. Acenetobacter sp. Mycobacterium sp. Candida sp. (yeast) Arthobacter sp.
Chlorinated solvents	Methylobacter sp. Methylococcus sp. (coccus) (methane users) Pseudomonas putida (phenol users)
Aromatics	Pseudomonas sp.
Chlorobenzoates	Nocardia sp. Arthrobacter sp.

Effluent treatment involves aerobic or anaerobic treatment of waste containing glycol ethers, polyurethane. Biodegradation of chlorophenol in contaminated soil could be achieved to the extent of 90%. White rot fungus *phanerochaete chrysosporium* can oxidise tributylene oxide with a powerful peroxidase and can be used to strip the paint from metal surface.

Production of ecofriendly paint additives

Microbial metabolites are being considered as the biodegradable environment friendly additives in paints. A novel concept on prevention of fouling was developed on immobilisation of a marine gram negative bacteria on hydrogels, which produced at least 2 extracellular compounds, inhibiting the settlement of invertibrate larvae on the surface, Xanthan, produced by *Xanthomonas compesris* finds application as highly effective rheological modifier in water based paints. Other biopolymers such as sceroglucan, pullulan, carageenan and alginate also have excellent potential. Certain UV stable pigments produced by *Serratia marcescens* has shown excellent potential. Certain UV stable pigments produced by *Serratia marcescens* has shown good promise. Biosurfactants also find applications in paints and protective coatings antibiotics produced by certain fungi can be used as antibouling agents in paints.

Future scope for microbes in paint industry

As evident from the above discussion, many common frontiers for interdisciplinary research are being identified and pursued. There is lot of scope for the future work in the following aspects:

- Development ecofriendly antifouling agents with low leaching rate and high effectivity.
- Isolation and screening of many more useful microorganisms including anaerobes, yeasts etc.
- Use of genetically modified microorganisms.
- Production and application of microbial enzymes.
- Technology to produce useful microbial metabodies as additives.

Table 5.9: Stages in development of biotechnological process in paint industry

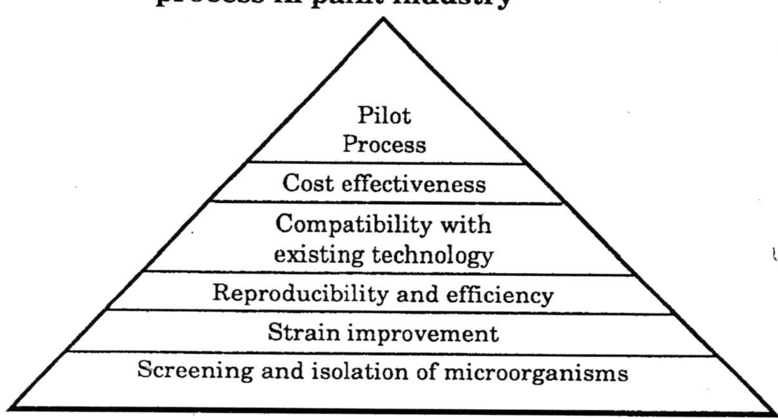

The stages in successful implementation of a biotechnological process for the industry are summarised in Table 5.9. Recently, microbiologists at the University of Bayreuth in Germany have developed a unique paint containing a bacterium *Oligotropha carboxidovorans* which feeds on carbon monoxide and converts it to carbon dioxide. This is an indication of the continuing research efforts being conducted in this field and also is a pointer to the paints of future. As biotechnology crosses new frontiers, the scope for its application in paint industry will also expand.

CHAPTER

6

APPLICATION OF PCR IN DETECTION AND MONITORING OF FUNGAL PLANT PATHOGENS

D.K. Arora, Tanuja Singh and Alok Srivastava

Polymerase chain reaction (PCR) is a powerful technique for directly amplifying short segments of the genome. This enzymatic reaction is so sensitive that a single DNA molecule can be amplified from the complex mixture of the genmcic sequences, which can generate large quantities of target DNA, and any nucleic acid sequences can be cloned, analyzed or modified. PCR procedure, from the standard method to its modifications and improvements, has greatly simplified DNA technologies. In mycology, many applications have already been described in taxonomy and population studies. However, PCR applications are being developed in many other fungal studies, including genetic analysis and the analysis of fungal-host interactions in both medical and plant pathology.

Because of its specificity and sensitivity, PCR is an attractive method for the detection of fungi and can be used to detect groups of strains, pathotypes, species or higher taxa. The major PCR-based techniques that have been considered at the species level for fungi are those developed from the rRNA gene cluster and the random amplification of polymorphic DNA. These methods include using restriction

Laboratory of Applied Mycology, Department of Botany, Banaras Hindu University, Varanasi-221 005, India.

fragment length polymorphism (RFLP) of mitochondrial and nuclear DNA and pattern differentiation of repetitive element. The advent of the PCR has expedited the molecular analysis of fungal genomes for both phylogenetic and population structure studies. Most of phylogenetic analyses are focused on the nuclear rDNA, which contains tandem repeats of three rRNA genes (28S, 18S and 5.8S) DNA sequence comparisons of the highly conserved nuclear 18S rRNA gene have been used to infer evolutionary relationships between different groups of fungi. The DNA sequenced that encodes ribosomal RNAs has been extensively used to study to taxonomic relationship and genetic variations in fungi. Ribosomal RNA gene cluster is found in both nuclei and mitochondria, and consists of highly conserved and valuable regions, which include the gene for the small 5.8s and large rRNA subunits. The conserved sequences found in the large subunit and small subunit genes have been exploited to study the many relationships among distantly related fungi. The spacer region between the subunits called internal transcribed spacer (ITS), and between the gene clusters, called the intergenic spacers (IGS), are considerably more variable than the subunit sequences, and have been widely used in studies on the relationships among species within a single genus or among intraspecific. The ITS region particularly, useful for molecular characterization studies in fungi for four main reasons: (i) The ITS region is relatively short (500-800 bp) and can easily be amplified by PCR, (ii) The multi-copy nature of the rDNA repeat makes the ITS region easy to amplify even from small dilute or degraded DNA samples, (iii) ITS region is also variable and therefore ITS generated RFLP restriction data can be used to develop rapid procedures for the identification of the fungal species and (iv) PCR-generated ITS species-specific probes can be produced quickly, without the need to produce a chromosomal library.

Many workers have selected sequences form the ITS region to develop species-specific probes because the sequences occur in multiple copies and tend to be similar within and between fungal species. rDNA-ITS region have been examined from a wide range of fungi and one of the most common taxonomic uses of the information obtained has been the development of species-specific probes and primers where restriction pattern form amplified ITS and SSU products have been used to differentiate species e.g., to develop a species-specific probe for *Pythium ultimum* for routine identification, species of *Phytophthora, Tuber* species, and *Leptosphacing maculans,. Colletotrichum* etc. At the intraspecifics level, differentiation of closely related fungal strains can be achieved by comparing more variable DNA regions such as the ribosomal IGS sequences, for which PCR primers have also been designed. Mitochondrial rDNA can also be easily analyzcd among fungi after amplification with consensus primers. Random PCR approaches are being increasingly used to generate molecular markers which are useful for taxonomy and for characterizing the fungal populations. The main advantage of these approaches is that previous knowledge of DNA sequence is not required so that any random primer can be tested to amplify and fungal DNA. In general, RAPD-PCR has been used most widely to define fungal population of species, intraspecific race and strain levels such as anastomosis groups in *Rhizoctonia solani* and pathogen groups. Some examples of RAPD-PCR at species level include the production of species-specific probes and primers from RAPD for *Fusarium oxysporum* f. sp. *dianthi, Phytophthora cinnamomi, Tuber magnatum* and *Glomus mossea.* Another application of RAPD-PCR has been used in the determination of individual strains within a particular population, some examples being toxin-producing strains of *Aspergillus Flavus*, and in strain authentication in species of *Trichoderma.* From the amplified products of RAPD markers, a SCAR (sequence characterized amplified regions) that can be identified by PCR—amplification using a pair of specific oligonucleotied primers. Other PCR-based methods are intermediate between random PCR approaches and target—directed amplification. Primers can be directed against microsatellites or minisatellites regions of fungal chromosomal DNA which are tandemly repeated motifs of 2-10 bp, or 15-30 bp, respectively. In most applications these primers have given similar levels of specificity to those seen with RAPD, and so results have been used to group fungi at intraspecific levels. However, in some instances microsatellite-primed PCR has been used to generate species-specific patterns. Further repetitive sequences have been examined for the use in fungal systematics, including primers derived from the M13 bacteriophage internal repeat, and the bacterial repetitive extragenic palindromic (REP) and enterobacterial repetitive intergenic consensus (ERIC) sequence. The recent development of molecular techniques has enabled the derivation of fungal phylogenies based on the analyses of proteins or nucleic acids, e.g., isozyme analysis, DND-DNA hybridization, electrophoretic karyotyping, RFLP and DNA sequencing. However, these molecular techniques have been difficult to apply to obligate parasites from which only small amounts of fungal materials are available. The PCR-based DNA sequence analysis has become the most useful method for inferring phylogenetic relationship between organisms. One can also observe the sequence variation, e.g. whether a change is transversion on transition, silent or selected, measure the degree of nucleotide bias, and the results allow the confirmation and their application to other taxa without the need to obtain strains or clones. The nuclear rRNA genes show little sequence divergence between closely related species and are useful for phylogenetic studies among distantly related organisms. Studies of the secondary structure of the rRNA have shown that eukaryotic large subunit rRNAs are compared of a conserved 'core' that has similar secondary structure to prokaryotic rRNAs, and interspersed 'divergent domains' that appear to have no prokaryotic homologue. DNA sequences of the divergent domains of the nuclear large subunit rDNA are useful for determining phylogenies of relatively closely related organisms. Within each repeat, the conserved regions are separated by two ITS regions, which show high divergence. Protein sequences also lend themselves to differential weighting of bases by codon position, and third position sites can provide a relatively good estimate of the natural substitution rate. Elongation factor-1a protein genes acting genes, B-tubulin genes, glyceraldehyde-3 phosphate dehydrogenase genes, chitin synthase genes and orotidine 5' monophosphate decarboxylase genes have been used for phylogenetic analyses fo eukaryotes including fungi. However, none of these, with a variety or organism. PCR-RFLP is now widely used for both fungal phylogeny and taxonomy. ITS/ IGS regions have mainly been used for PCR-RFLP as these regions are highly polymorphic and can easily amplified by PCR using universal primers.

Every food or feed commodity can be affected by fungus, and about 25% of the production of plant-derived foods are spoiled by many food borne fungi, capable of producing mycotoxins which are hazardous to both the health of human beings as well as animals. These mycotoxins belong to different classes, like polyketide, isoprene and amino-acid substances. Due to different molecular structures of these mycotoxins, their effect on health of human beings and animals also varies e.g. teratogenic, immunosuppressive, nephrotoxic hepatotoxic to carcinogenic. Therefore, to avoid this situation, early detection of these fungi are important to control the mycological status of food and feed commodities. These problems can be overcome by the use of the PCR for the rapid detection of mcotoxinogenic, which can reduce the detection time from several days to several hours. However, they can only give information about the presence of potential mycotoxin productng fungus in a sample but can not detect the mycotoxin itself. If an appropriate target sequence is chosen, the method can differentiate between mycotoxin producing and non-producing strains of a species. The most direct procedure for the development of a diagnostic PCR method for mycotoxin producing fungi is targeting the mycotoxin biosynthetic genes, because these target sequences should not be present in strains of the same species that are not able to produce mycotoxins. However, to date, only some of the genes of mycotoxin aflatoxins, sterigmatocystin, trichothecene, fumonisin, polykatide synthesis and biosynthetic pathways have been cloned and sequenced. Polykatide synthesis are often involved in fungal secondary metabolism and can be found in various species and if their sequences are not available, other opportunities for the identification of target sequences exist, such as the determination of specificities in ITS sequences/random sequences pre-selected by RAPDs from structural genes may also serve as target sites.

In attempting to overcome low sensitivity, poor specificity and delay, molecular approaches to diagnosis of fungal infection have been adopted. Much research work has been made into the development of PCR as a diagnostic tool in the field of medical mycology. The major developments in PCR-based protocol for the diagnosis of invasive candidosis, aspergillosis, cryptococcosis and other infections due to dimorphic fungi. Commercially produced PCR-based protocols are now widely available for the diagnosis of various infectious of various common fungal pathogens have been described. Now commercial biochemical tests are available for identifying clinical yeast isolates, but problem of non-specificity have been encountered with more unusual species. If appropriate therapy is to improve the prognosis of the immunocompromised patient with systematic fungal infection, early diagnosis is required. When developing PCR based diagnostic method various factors like selection of appropriate target sequence for amplification, sample choice and its preparation and display of maplicons resulting from PCR can be adopted to optimize both sensitivity and specificity. The specificity is determined by degree of homology shared between the target sequence and DNA sequences of different genera or species (e.g. a highly conserved region from a gene universally shared by all fungi is a suitable target sequence for identification of whether an infection is bacterial or fungal in origin). This strategy has been adopted by several groups who have selected highly conserved region within fungal ribosomal RNA (r-RNA) genes as target sequences for amplification. Ideally, the presence of target DNA in the chosen specimen will be clinically significant; the detection of *Candida* infections involve the examination of blood samples as detection of the fungus in blood is regarded as clinically significant. Another approach to amplicon analysis is to look to single-strand conformational polymorphism (SSCP). Minor sequence changes in highly conserved regions of DNA will cause subtle morphological changes in DNA tertiary structure resulting changes in fragment's mobility differences of only one base pair can be resolved by separating the DNA products electrophoretically under non-denaturing conditions on a high—resolution acrylamide gel, allowing discrimination between species and even strain. A universal PCR based on the 18S rRNA gene combined with SSCP detection to identify a variety of opportunistic fungal pathogens has also been developed. The SSCP gel patterns enabled discrimination between *Candida* spp., *Aspergillus* spp., C.*neoformans, Pseudallescheria boydii* and R. *arrbizus.*

PCR is now widely used to detect and identify microorganisms, have made a large impact on research in the plant-fungal interactions, including both plant pathogens and symbionts on or in their hosts more rapidly and reliably than before and for a few fungi these methods are quantitative. The use of PCR for quantification of fungal biomass and infected tissue is not straightforward. This is only because of PCR includes an exponential phase, so even a minute differences in reaction can give rise dramatic difference of final product therefore, internal control DNA, which is amplified in the same PCR reaction, should be included. Quantitative PCR has been used in the detection of Mycorrhizal fungi, *Verticillium* spp., and *Microdochium nivale.* PCR has also simplified the isolation of genes involved in plant fungal interactions in which some genes involved in fungal infection and plant responses to infection are common. Many host-pathogen combinations and some of these have been characterized at the nucleotide or amino acid level. Once a protein or nucleotide sequence is available for one gene, primers can be designed to allow related genes to be isolated very quickly. This approach usually involves the use of degenerate primers because even within conserved regions of genes there is often some variation, e.g. between different plants or fungi or between different members of a gene family in the same plant. However, PCR can also be used to isolate genes involved in plant fungal interactions without prior knowledge of the protein involved. These methods detect differences at the RNA level that RNA level that correlate with fungal infection of differences at the DNA level that correlate with a trait such as plant resistance or fungal pathogenicity. There are various uses for the genes once isolated, in addition to characterizing the gene structure and organization. PCR can also be used for *in vitro* mutagenesis to study the effects particular amino acid residues on the biological activity of the gene product. Altered primers can be used to replace the nucleotide coding for single or multiple amino acid residues in the protein. The ability to monitor the amounts of different fungi infecting the same plant at the same time by quantitative PCR will be of great value in studying disease complexes and interactions between fungi, e.g. synergism and competition. This could be an area for development in the future, not just for detection of the fungi themselves but also for the detection of mRNA from expressed in plant—fungal interactions.

CHAPTER

7

OPTIMISATION OF POLYMERASE CHAIN REACTION (PCR) FOR THE DETECTION OF CYTOMEGALOVIRUS (CMV) IN URINE

A. Pultoo, H. Jankee, G. Meetoo*, M.N. Pyndiah* and G. Khittoo*

Cytomegalovirus (CMV) is a human viral pathogen belonging to the Herpes Virus family (9). There are several mechanisms of CMV transmission including congenitally acquired infection in neonates, sexual contacts, breast-feeding, saliva and blood transfusions (1). This virus has a worldwide distribution with a prevalence varying between 40%-100% and is more common in developing countries. Infections with CMV are usually asymptomatic which often give rise to undetected latent infections and reinfections (3). Moreover (CMV) is an important cause of mortality and morbidity in immunocompromised individuals such as newborns AIDS patients and transplant recipients (5,8,10).

Recognition of infection by CMV is currently made by detecting the virus in cultured fibroblasts or demonstrating the presence of circulating CMV antibodies in individuals. Both approaches, however, have significant disadvantages. The tissue culture procedures are elaborate, time consuming and considered by some to lack sensitivity (2,11). Detection

Faculty of Science, University of Mauritius, Reduit, Mauritius
*Virology laboratory, Victoria hospital, Candos, Mauritius

of specific antibody response can only give indirect evidence of CMV infection. On the other hand the polymerase chain reaction with its potential speed, specificity and sensitivity has been shown to be very useful in several medical diagnosis (4,6). In this work, we optimised PCR reaction conditions and compared its efficiency with that of cell culture for detection of CMV in urine of hearing-impaired and mentally retarded children.

Collection of Specimens

Samples for PCR and cell culture were taken from 71 mentally retarded children and 28 deaf children aged between 5 and 12 years. Freshly voided urine samples were collected from the children in sterile 200 ml plastic containers. Immediately after collection, 3 ml of each urine samples were placed in a 10 ml sterile bijou vial containing 3 ml CMV transport medium (Eagles MEM, 4% calf serum, 35% [w/v] sorbitol, 100 μg/ml streptomycin, 100 U/ml penicillin and 2.5 μg/ml fungizone) and put on ice to maintain viral activity. Another 7 ml of urine was placed in sterile 1.5 ml plastic tubes for PEG precipitation and polymerase chain reaction.

Virus isolation by cell culture

Attempts were made to isolate CMV from urine specimens by using locally prepared low passage human embryonic lung fibroblasts (HEL) monolayers maintained in confluent growth in closed borosilicate tubes. Inoculations were done in duplicates using 0.2 and 0.3 ml of urine. Cultures were maintained at 37°C for 4 weeks because of slow growth of CMV. The maintenance media containing Eagle's minimum essential medium (MEM), 2% inactivated calf serum, 0.1% sodium bicarbonate, 2 mM L-glutamine and antibiotics (100 μg/ml streptomycin. 100 U/ml penicillin and 2.5 μg/ml fungizone) were changed twice a week. Inoculated cultures were examined twice a week for characteristic cytopathic effect of CMV distinguished by focal formation of giant cells growing in line with the cell sheet. Positive cultures were passaged a second time to confirm isolation of CMV.

Polymerase chain reaction for detection of CMV DNA in urine samples

PEG precipitation

Polyethylene glycol (PEG) precipitation was carried out in order to concentrate the virus from a large volume of urine and to remove PCR inhibitors in the urine. Seven ml of urine was mixed with 7ml of PEG 6000 in Tris pH 7.4 buffer (0.05 ml Tris, 10 mM $MgCl_2$, 0.35 M NaCl) and left overnight at 4°C. The optimum concentration of PEG and NaCl were determined by Yamamoto *et al.* (13). The precipitate formed was dissolved in 200 µl of Tris pH 7.4.

Optimisation of polymerase chain reaction

The requirement of an optimal PCR reaction is to amplify a specific locus of a DNA (which may be found in very low amounts) without any specific by-products. Therefore optimisation of the reaction conditions i.e. incubation time and temperatures, number of cycles, and a concentration of $MgCl_2$, dNTP and primers to obtain a specific and sensitive PCR is mandatory.

Primer concentration

Primers that closely matched in Tm, and spanning a region of 240 base pairs, which mapped to the perfectly conserved immediate early gene region 2 of HCMV strain Towne were selected (11).

CMV A : 5′ CCC GAC TTT ACC ATC CAG TA 3′ (sense)

CMV B : 5′ AAG ACG AAG AGG AAC TAT CT 3′ (antisense)

Primers concentrations in the range between 10 and 100 pmoles were assayed.

$MgCl_2$ and dNTP concentrations

The two most critical components in the PCR are the dNTP/$MgCl_2$ concentrations and the annealing temperature. The $MgCl_2$ and dNTP concentrations are dependent on each other, as well as on the concentration of primers (6). To work properly, Taq DNA polymerase requires free $MgCl_2$ (besides the magnesium bound by the dNTP and the DNA). Different concentrations of dNTP and $MgCl_2$ were assayed simultaneously in order to optimise the PCR.

Incubation time and temperatures

The Tm of primers was 60°C. Annealing needs to take place at a sufficiently high temperature to allow only the perfect DNA-DNA matches to occur in the reaction because Taq polymerase has reduced activity at low temperatures (Optimum temperature for Taq polymerase is 72°). Therefore, to avoid non-specific amplification and parasite bands the annealing temperature was raised to 60°C. In general, shorter time allowed for annealing improves the specificity of PCR (6); consequently the annealing time was reduced to 30 seconds. The denaturation and extension time used were 1 min and 2 min respectively so as to allow sufficient time for denaturation of target DNA and complete extension of the primers. Too long a denaturation time would have increased the time the DNA polymerase had been subjected to high temperatures, increasing the percentage of polymerase that lose their activity. Very long amplification time would have increased the likelihood of unspecific amplification products.

Number of cycles

Increasing the number of cycles may enhance an anaemic reaction, but this modification can also lead to the generation of spurious bands and to smears composed of high-molecular weight products rich in single-stranded DNA. Moreover after 33-35 cycles, the accumulation of PCr product is no longer linear, but reaches a plateau. Therefore a 30 cycles PCR was used.

Sensitivity of PCR

Serial 10-fold dilutions of DNA from HEL infected with CMV were carried out until an electrophoretic band could not be observed after PCR, in order to define the sensitivity and specificity of the PCR.

Amplification

For PCR, 5 μl of the product from PEG precipitation was added to a 45 μl reaction mixture containing 10 mM tris. HCI pH 9.6, 2.5 mM $MgCl_2$, 0.2 μg/ul bovine serum albumin, 50 pmloes of each primers A and B, 0.2 mM dNTP's, 1.0 U of Taq DNA polymerase and overlaid with μl of light mineral oil.

Agarose gel electrophoresis

Five μl of each amplified mixture was added to 5 μl of 2x loading dye containing 60% (v/v) glycerol and 0.15 each of bromophenol blue and xylene cyanole in TAE buffer (40 mM Tris base, 12 mM disodium EDTA [pH 8.0], 0.01% (v/v) glacial acetic acid) and loaded on 1% agarose (Sigma). To identify the bands, standard molecular weight marker pBR 322Hae III digest (Sigma) or 174 Hae III digest (Amersham pharmacia biotech) were run in parallel. The gel was run in TAE buffer for 45 min at 100 V on a minigel (Fisons), stained with 0.1 μg/ml ethidium bromide, visualized under ultraviolet transilluminator and documented by an imaging system (Advanced American Biotechnology, U.S.A.).

Results

Optimisation of $MgCl_2$ and dNTP concentrations

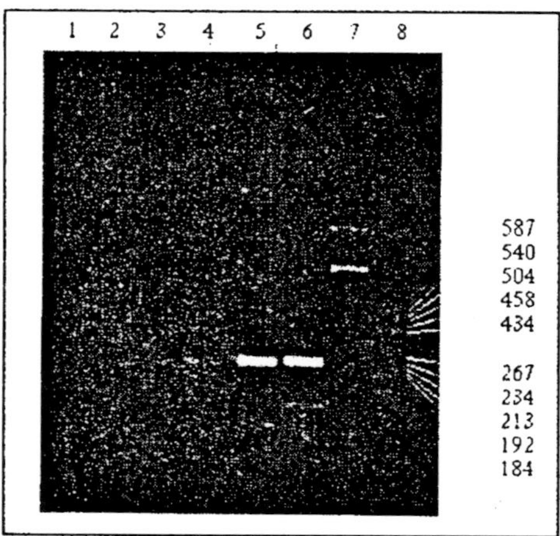

Fig. 7.1: Optimisation of primers concentration using 10ng of DNA from HEL cells infected with CMV and 2mm $MgCl_2$ Lanes 1-7: 0.1 pm, 0.3 pm, 0.4 pm, 0.5 pm, 0.6 pm, 0.7 pm, respectively; lane 8: molecular weight marker-pBR322 Hae III digest. Optimum concentration of primers were 50 pmoles (lane 5)

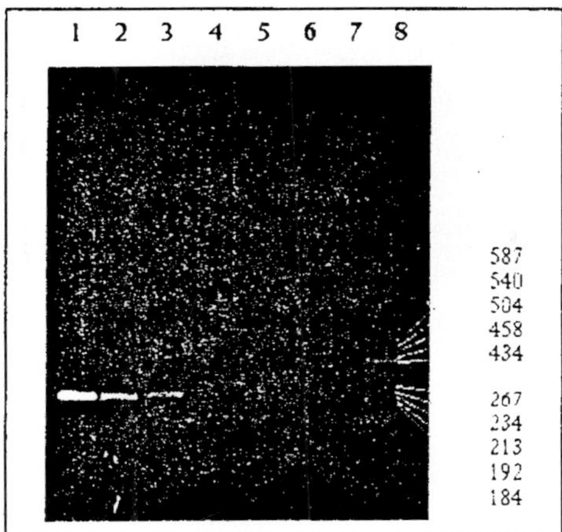

Fig. 7.2: Sensitivity of PCR applied to dilutions of 245 μg of DNA from HEL cells infected with CMV; Lanes 1-7: Serial DNA dilutions 10^{-1}, 10^{-2}, 10^{-3}, 10^{-4}, 10^{-5}, 10^{-6}, 10^{-7}, respectively. Lowest detection limit of PCR was 24.5 pg of DNA from HEL cells infected with CMV (lane 7). Lane 8: molecular weight market-PBR 322 Hae III digest (Sigma)

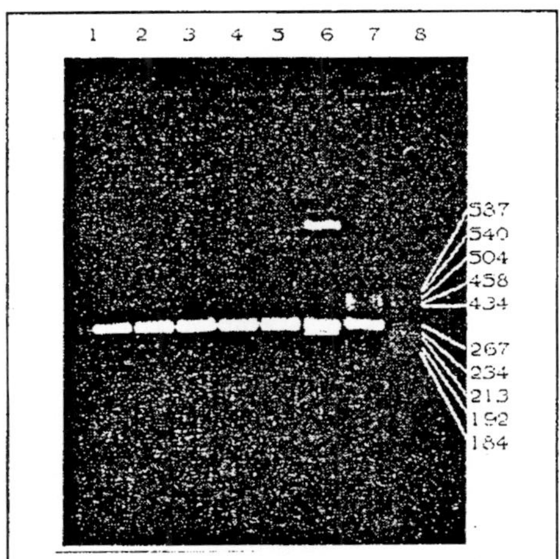

Fig. 7.3: Optimisation of $MgCl_2$ Concentrations with 10ng of DNA from HEL infected with CMV. Lane 1-7: $MgCl_2$ concentrations-0.5 mM, 1mM, 1.5 mM, 2.0 mM, 2.5 mM, 3.0 mM, and 3.5 mM respectively. Lane 8: marker pBR322 Hae II digest. Highest efficiency was obtained with 2.5 mM $MgCl_2$.

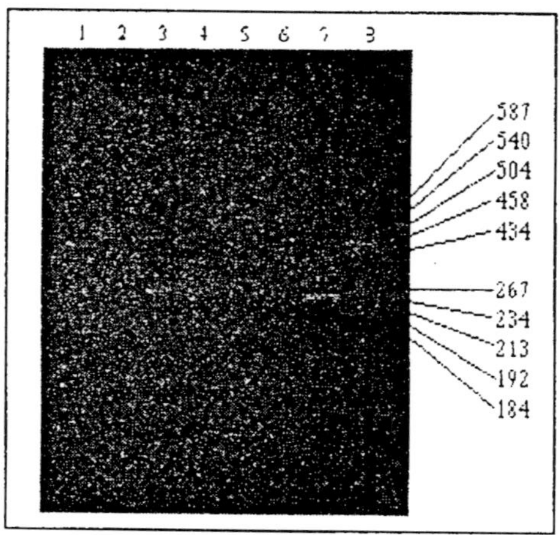

Fig. 7.4: Agarose gel electrophoresis of products of PCR performed on urine samples from deaf children; lanes 1-2 negative urine samples; Lane 3-4: positive urine sample; lane 5: control (no enzyme); lane 6: control (n DNA); lane 7: Positive control; lane 8: pBR 322 Hae III marker.

By combining various amounts of $MgCI_2$ and dNTP, it was found that 0.2 mM each dNTP worked well with 2.5 mM $MgCl_2$. Lower levels of $MgCl_2$ resulted in a decrease in the efficiency of amplification and reduction in the dNTP concentration leads to the formation of primer artefacts as a consequence of excess free Mg2+resulting from the lowered dNTP concentration. Lower levels of dNTP also lead to inconsistent amplification results and frequently the reactions produced only a subset of the targeted amplicons. High levels of $MgCl_2$ resulted in unspecific amplification and with high concentrations of dNTP amplification was inhibited due to reduction in free $MgCl_2$.

Optimisation of primers concentrations

The PCR was inefficient when the concentration of primers was limiting. Conversely, the presence of higher concentration of primers causes annealing at ectopic sites, with consequent amplification of undesirable non-target sequences. Optimum concentrations of primers were found to be 50 pmoles.

Comparison of PCR and cell culture

A total of 99 urine samples from mentally retarded or hearing-impaired children were tested by both cell culture and PCR. All urine samples (25) positive by cell culture were PCR-positive. Five samples were cell culture-negative but PCR-positive. None of the samples positive by cell culture was PCR-negative. Sixty-nine samples were negative by both cell culture and PCR. Hence, compared to cell culture, urine PCR had a sensitivity of 100% and a specificity of 93%.

Table 7.1: Comparison of PCR and cell culture

	Cell Culture +	*Cell Culture –*
PCR–	25	5
PCR-	0	69

Compared to cell culture, urine PCR had a sensitivity of 100% and a specificity of 93%.

Flase-negative results by cell culture may be due to effect of low viral load. PCR was very sensitive because it can detect low viral load and in addition PEG precipitation concentrated the virus in a large volume of urine as well as removing PCR inhibitors in the urine.

Using the PCR to generate large amounts of a desired product is a double-edged sword. Failure to amplify at optimum conditions leads to absence of any product or presence of non-specific amplicons. Contamination may be another major hindrance. The smallest amount of contaminating DNA may be amplified, resulting in misleading or ambiguous results. Hence scrupulous care must be maintained whenever PCR work is undertaken. (6).

The development of an efficient PCR required strategic planning and multiple attempts to optimise reaction conditions. A sensitive and specific PCR was established by careful choice of primers, optimisation of the concentrations of $MgCl_2$, dNTPs and primers, and adjustment of incubation time and temperatures. The high specificity of our PCR was confirmed by generation of only the intended 240 bp product, and its high sensitivity by amplification of 24.5 pg of HEL infected with CMV, which is comparable to other reports (7, 12). Reaction mixtures without DNA or DNA from uninfected HEL cells gave negative results in each assay run, confirming the efficiency of measures preventing contamination.

The successful implementation of PCR DNA amplification procedures to detect CMV in the urine of infected children should facilitate a positive diagnosis of CMV that would not necessitate prior isolation and purification of the virus or its DNA and may be used to diagnose a variety of other CMV related diseases. Given the dependability of PCR and its rapidity (it can be completed in less than 6 hrs), kit may be utilised for monitoring of disease progression and antiviral therapy (2,3).

References

1. Chopra, H.L. 1985, "Textbook of Medical Microbiology," New Delhi, Seema Publications. pp. 632-633.
2. Chou, S., and Scott, K.M. 1988. "Rapid quantitation of cytomegalovirus and assay of neutralising antibody by using monoclonal antibody to the major immediate early protein. J. Clin Microbiol 26: pp. 504-507.
3. Dejong, M.D., Galasso, G.J., Gazzard, B., Griffiths, P.D., Jabs, D.A., Kern, E.A., and Spector, S.A. 1998. "Summary of the II international symposium on cytomegalovirus. Antiviral Research 39. pp. 141-162.
4. Demmler, G.J., Buffone, G.J., Schimber, C.M. and May, R.A. 1988. "Detection of cytomegalovirus in urine from newborns by using polymerase chain reaction DNA amplification. J infect Dis 158: pp. 1177-1184.
5. De Schryver, A., De Backer, G., and Van Renterghen, L. 1998. "Cytomegalovirus infection: epidemiology and association with congenital malformations. Arch Public Health 56: pp. 233-250.
6. Dieffenbach, C., and Dveksler, G.S. 1995. "PCR Primer: A laboratory manual. Cold harbour Laboratory, New York.
7. Einsele, H., Vallbracht, A., Jahn, G., Kandolf, R., and Muller, C.A. 1998. "Hybridisation techniques provide improved sensitivity for HCMV detection and allow quantitation of the virus in clinical samples. J Virol Methods 26: pp. 91-99.

8 Kaariainen, L. 1966. "Cytomegalovirus mononucleosis: Isolation of a virus and demonstration of subclinical infections after fresh blood transfusion in connection with open heart surgery. Am Med Exp Fen. 44. pp. 297-301.

9. Murphy, F.A., Fauquet, C.M., Bishop, D.H.L., Ghabial, S.A., Jarvis, A.W., Martelli, G.P., Mayo, M.A., and Summers, M.D. 1995. "Virus taxonomy. Sixth Report of the International Committee on Taxonomy of Viruses. Springer-Wien.

10. Spector, S.A., Weingeist, T., and Pollard, R.B. 1993. "A randomized, controlled study of intravenous ganciclovir therapy for cytomegalovirus peripheral retinitis in Aids. J Infect Dist. 168: 557-563.

11. Sternberg, R.M., Witte, P.R., and Stinski, M.F. 1985. "Multiple spliced and unspliced transcripts from human cytomegalovirus immediate-early region 2 and evidence for a common initiator site within immediate-early gene region 1. Journal of Virology 56: pp. 665-675.

12. Yamaguchi, Y., Hironaka, T., Kajiwara, M., Tateno, E., Kita, H., and Hiraj, K. 1992. "Increased sensitivity for detection of human cytomegalovirus in urine by removal of inhibitors for the polymerase chain reaction. J Virol Methods 37: pp. 209-215.

13. Yamamoto, K.R., Alberts, B.N., Benzinger, R., Lowhorne, L., and Triener, G. 1970. "Rapid bacteriophage sedimentation in the presence of polythylene glycol and its application to large-scale virus purification. Virology 40: pp. 734-744.

14. Zukermann, A.J., Banatvala, J.E. and Pattison, J.R. 1995. "Principles and Practice of Clinical Virology (3rd ed.), pp. 68-108, West Sussex, John Wiley and Sons Ltd.

CHAPTER

8

PROSPECTIVES OF PLANT BIOTECHNOLOGY IN AGRICULTURE

K.R. Koundal

Recently, biotechnological approaches have become available for genetic variation. Culture of plant cells *in vitro* generates genetic variation, called somaclonal variation from which several useful somaclones have been isolated and released as commercial varieties. Fusion of protoplasts of selected plant species constitutes somatic hybridization which often yields hybrid plants between sexually incompatible species. Anther culture yields haploid plants, chromosome developing of such plants generates homozygous lines in a very short time. Several commercial varieties have been developed by combining anther culture with the conventional breeding programme.

Classical plant breeding is limited to the introduction of required characters into plant by genetic crossing during sexual reproduction. This classical approach is useful if the desired genetic improvement is encoded by single gene. If a trait is determined by several genes, it may be extremely difficult to introduce all the genes responsible for crop improvement by using conventional breeding techniques. Moreover, we can cross only two sexually compatible plant

Project Director, National Research Centre on Plant Biotechnology, IARI, New Delhi-110012, India.

species and it takes 10-15 years to introduce a new variety. It is in this text that the newly emerging area of plant biotechnology, viz. Plant genetic engineering, assumed greater significance. Using plant biotechnology, plant can be genetically modified by manipulating plants own genes and also by introducing genes from taxonomically unrelated plants and other organisms, such as fungi, viruses and even animals. Principally, plant biotechnology approaches could be grouped into two categories-I. Plant tissue culture and II. Plant genetic engineering.

I. PLANT TISSUE CULTURE

Recent advances in tissue culture technology have made it possible to culture and regenerate whole plants from cells, protoplasts and tissue *in vitro* in a medium containing carbon source mineral growth factors and growth regulators (1) The plant regenerated from callus a resuspension culture often shown an increased of genetic variation which is of tremendous importance in breeding for crop improvement. Some of the areas where tissue culture has been applied with success are as follows:

1. Micropropagation

Micropropagation is the term used for propagation in the laboratory. Under controlled and aspeptic conditions, rather than in greenhouse a nursery Micropropagation can be armed and by using (i) shoot tip culture (ii) Culture of nods (iii) micropropagation by organogenesis and somatic embryogenesis. Plant developed under these conditions are transferred to green house for acclimatization and thereafter used as manual plants. Micropropagation is a multi billion dollar industry being practised in several small and large nurseries and commercial labs all over the world. The successful examples include, high value ornamentals and to large plants such as chrysanthemum, lilies, gladioli rose, carnation, papaya and strawberry etc. In plantation and spice crops tissue culture techniques have been successfully employed to increase productivity. Successful examples include banana, cardamom and black pepper.

2. Somaclonal variation

The variability generated by the use of tissue culture is termed as somaclonal variation. In several plant species variation has been documented for agronomic traits such as disease resistance, plant height yield maturity quality and plant type. Somaclonal variation has resulted in the improvement of cultivars in sugarcane tomato celery, rice sorghum and apple. In tomato promising cultivars 'DNAP-9' and 'DNAP-17' have been developed which had high solid content and *fusarium* resistance. In brassica, a variety called 'Pusa Jai Kisan" has been developed with improved yield of oil content. Similarly in banana cavendish dwarf somaclonal resistant to *Fusarium oxysporium cubens* has been achieved. Somaclonal variation have sexual advantages mainly that it is cheap and needs limited tissue culture facilities.

3. Embryo Rescue

Embryo rescue is a technique in which immature zygotene embryos are dissected out from the developing fruits (seeds) and grow in suitable medium that either callus the embryos to grow directly into plantlets or allows differentiation of embryo tissue into shoots and roots.

Different cross may fail due to one or more reasons. When embryo fail to develop due to endosperm degeneration, embryo culture is used to recover hybrid plants. This is called hybrid rescue. Embryo culture has been widely used for this purpose. Some recent examples are the recovery of hybrids from *Hardeum vulgare x secale cereale, H. vulgare x Triticum* species including *T. aestivum* etc. In addition, some interspecific crosses of *H. vulgare x H. bulbusum,* wheat x maize yielded haploid embryos. Embryo culture is used to recover the haploid plants from such crosses which are then used in breeding programmes.

4. Production of haploid plants

Haploids are of great value in agriculture. They serve the important purpose. Direct screening for recessive mutations in plants is not possible since plant cells are diploid or polyploid. Generation of haploid plants or cell lines would be particularly useful for screening for recessive mutations. The second important use of the haploid plants is the scope it offers to develop homozygous diploid plants following chromosoms breeding cycle through anther culture is very useful in several crops. Haploid have been effectively coupled with breeding programmes of crops like rice, wheat, barley, tobacco etc., this has resulted in the development of several varieties. In China, over 100 rice varieties were developed using anther culture route, many of them represent yield enhancement of 10% to 25% and were cultivated in substantially large areas.

5. Somatic hybridization

Somatic hybridzation involves physical union or fusion of protoplasts from two parents. A wide variety of cell fusion hybrid plants have been developed (2). However, plant regeneration remains one of the principal obstacle for the extensive use of cell fusion hybridization for general application in plant breeding. Cybrid (fusiỏn of a protoplast with an enucleated protoplast) have been developed in rapeseed using somatic cell hybridization technique in combination with traditional plant breeding. Using this technology, it is possible to introduce gene across species barrier.

Several successful example include Arabido brassica (*Arabidopsis Thaliana* + *Brassica campestris*), tomatoes and potatoes (potato + tomato) and Erucobrassica (*Brassica napus* + *Enica sativa*) somatic hybrids of sexually incompatible species of citrus have also been produced. The second application of protoplast fusion is the production of cytoplasmic hybrids (cybrids) in which nuclear genome of one parent and the organelles genome of another parents are combined.

Transfer of cytoplasmic sterility (CMS) to breeding lines ensures the development of hybrid varieties. Protoplast fusion enables the transfer of cytoplasmic male sterility (CMS) between elite breeding lines. Successful examples include rice, petunia, Brassica, Nicotiana where CMS traits have been transferred to elite lines. The cybrids in these studies originated either spontaneously during protoplast fusion or even produced by either enucleation or X-irradiation of protoplasts of one of the fusion partners. This has been demonstrated in tomato by transfer of mitochondia from *Solanum Acaule* into tomato by protoplast fusion. Protoplast of tomato were treated with iodoacetate while those of *S. acaule* are irradiated with X-rays. The male sterility was inherited for several generations. In addition, and more importantly, restorers of these CMS (cytoplasmic male sterility) lines are available in tomato, restorers are essential for practical exploitation of CMS in hybrid seed production.

II. PLANT GENETIC ENGINEERING

Progress in plant genetic engineering has been dependent on efficient methods of introducing foreign DNA into plant cells. Gene transfer with plant cells can be achieved by either direct uptake of DNA or the natural process of gene transfer carried out by the soil bacterium *Agrobacterium* and more recently vectors based on the genomes of plant viruses have become available. In addition, the versatility of gene transfer vectors is such that they may be used to isolate genes not amenable to isolation using conventional protocols. The plant obtained through genetic engineering contain a gene or genes usually from an unrelated organism, such genes are called transgenes and the plant containing transgenes are known as transgenic plants (3). The real potential or genetic engineering to help address some of the most serious concerns of world agriculture has only recently begun to be exploited. The following examples show how genetic engineering can be applied to some of the specific problems of agriculture indicating the potential for benefits.

1. Improvement of Nutritional Quality

Seed storage proteins, which constitute a major source of proteins in the human diet and animal feed suffer from the disadvantage of having very low levels of essential amino acids, proteins of pulses are deficient in sulphur containing amino acids cysteine and methionine and cereal grains proteins are deficient in lysine and tryptophan (4). Genetic engineering approach involves the identification of seed storage proteins that have high levels of essential amino acids and then engineering their expression in the target plants. In one instance achimeric gene with a phaseolin (seed storage protein) gene promoter and a coding region of a methionine rich protein (2S albumin) from brazilnut was constructed and introduced into tobacco. The transgenic tobacco plants expressed the 2S protein in the seed tobacco plants expressed the 2S protein in the seed resulting in a 30% increase in the levels methionine. These experiments demonstrate the feasibility of improving the quality of seed storage proteins. Production of oils and carbohydrate with value addition have also been demonstrated in transgenic plants. Seed specific expression of ACP-desaturase antigene lead to a decrease in desaturase concentrations and a concomitant accumulation of stearole in rapeseed embryos. Starch synthesis has also been modified genetically. The gene for the initial unique embryo in starch biosynthesis. ADP-glucose pyrophosphorylase (ADPGPP) has been cloned from an *E. coli* mutant transferred to potato plants and expressed under the control of patatin promoter. This manipulation increased the starch content of tubers.

In addition research have introduced three new genes into rice. Two from *deffodils* and one from microorganism. The transgenic rice exhibits an increased production of beta carotene as a precursor of vitamin A and the seed is yellow in colour (5). Such yellow or golden rice may be useful to help treat the problem of vitamin A deficiency in young children living in the tropics. Similarly transgenic rice with elevated levels of iron has been produced using genes involved in the production of an iron-binding protein and in the production of an enzyme that facilitates iron availability in the human diet.

2. Herbicide Resistance

The main limitation is the specificity of the herbicide. There is no known herbicide that is toxic to weeds that is not also toxic to crop plants to some degree. Therefore, herbicide resistance can be conferred to on a crop plant by two strategies. Either by addition of a gene that allows the plant to detoxify the herbicide or addition of a gene that makes the target protein either in large amount or in resistant form.

Both of these strategies have been used in the past. In order to choose the most favourable strategy exact knowledge of the site of action and degradation pathway of herbicide is necessary. Resistance of atrazine can be induced in plants by a single gene point mutation is chloroplast encoded psb A gene. The mutated psb A gene could be transferred to sensitive plants to make them resistant. Glyphosate resistance has been developed in petunia by amplification of ESP synthase gene. Transgenic tobacco plants resistant to glyphosate have also been developed by introducing mutation in ESP synthase and making it insensitive to glyphosate. Plants having resistance to herbicide in several economical plants such as tobacco, tomato, oilseed rape, flax, soybean, cotton have been engineered and commercialized (6).

3. Virus Resistance

Infection of crop plants by viruses can lead to reduced yield lowered product quality and sometime to complete failure. Various protocols for engineering virus resistance in host plants have been designed including coat protein mediated resistance, expression of satellite RNA a replicase sequences, interference by defective RNA or DNA sequences and the use of antisense RNA. The coat protein gene from the TMV were expressed in the host plant (tobacco or tomato) confers high level protection from incoming virus. Researchers have used novel techniques that minimize 'genetic minimization' by creating transgenic rice plants that are resistant to rice yellow mottle virus RYMV. Resistant transgenic varieties are currently entering field trials to test the effectiveness of their resistance to RYMV (7). This could provide a solution to the threat of total crop failure in the sup Saharan Africa rice growing regions.

4. Resistance towards insects pests

Crop protection from insects plays a very important role in agricultural productivity. Progress in developing insect control in transgenic plants has been initially achieved through the expression of insect control protein genes of *Bacillus thuringiensis*. The insecticidal protein genes from Bt conferred for reaching resistance to classes of certain lepidoptera species when transgenic tomato, potato, maize and cotton plants were produced. There are several Bt genes encoding proteins that are largely insecticidal and interact both specific receptor molecules in insect midgut, make weaker and eventually kill the insect. The technical improvement have added the effectiveness of the endotoxin strategies. Firstly, the translational fusion for example cry 1A and cry 1C genes are superior to mild type endotoxin genes and should be engineered in the target plants. The major genetically modified by cry 1A(c) for commercial cultivation include cotton, potato and corn. Secondly fully synthetic endotoxin genes appropriately designed also confer better resistance in cups than have wild type. However, there are reports of insect population developing resistance against the toxin by mutations in the receptor protein genes. An obvious alternative is to transform the target plant with few or more different genes so that insect would have more than one endotoxin protein with which to cope and use promoter linked to endotoxin gene that are normally silent but become active after the pest insect attack (i.e. word inducible promoter).

Many plants have evolved a natural mechanism of defence against herbivorous insects by accumulating proteinase inhibitors (proteins which inhibit the activity of proteinase enzymes) at a concentrations that will cause metabolic inhibition upon ingestion by insects. Compared to Bt toxins, proteinsase inhibitors have a broader spectrum of metabolic inhibition and so can be used to control many types of insects (8) Transgenic tobacco plants expressing high levels of trypsin inhibitor protein encoded by a gene of an African variety of cowpea showed decreased insect damage by *Heliothis virescences,* tobacco bud worms, a pest of tobacco. It was also demonstrated that cowpea trypsin inhibitor killed a variety of depidopteran and coleopteran insects. Lectin have also been reported to impart resistance against aphids in many crops lectin from snow dwarf (Galanthus nivallis) is the first plant protein known to damage any suckers i.e. aphids of grasshoppers. This strategy of making transgenic plants is relatively safe and compatible with environmental protection.

5. Fungal Pathogen Resistance

Chitinase and B 1-3 glucanases belong to the so called pathogenesis related protein (PR proteins) because their synthesis increases markedly after attack of phytopathogens. They are shown to have antifungal activity. *Brassica napus* plants are tobacco plants engineered with bean chitinase showed significant resistance to *R. solani which* cause a severe disease in plants. A substantial resistance probably only can be engineered using combination of several genes (pyramiding). While genetic modification technology provides access to new gene pools for sources of resistance it needs to be established that these of resistance will be more stable than the traditional intra species sources. There is need therefore, for more research on transgenic plants that have been made resistant to local pests to access their sustainability in the face of increased selective presence for ever more virulent pests.

6. Extended Shelf-life of Fruits

Delay in ripening extends the shelf life and keeping quality of fruits thus long distance shipping of fruits with minimize damage. In tomato, plants have been engineered for reduced synthesis of polygalacturonase which is responsible for softening the cell wall. Using the antisense technology, antisense RNA for polygalacturonase has been inserted. When normal pGU mRNA is synthesized the antisense RNA binds to it and prevents synthesis of PGU. Antisense RNA against ACC synthase was also expressed in tomato plants and this inhibited ethylene biosynthesis by about 99.5% and markedly delayed fruit ripening. Genetically engineered tomatoes have become a reality of commercial company, collagen is set to market genetically engineered 'Flvr-Savr' tomatoes developed using antisense RNA technology. These tomatoes have an extended shelf life and increased resistance to brushing.

In many cases small-scale farmers suffer heavy losses due to excessive or unocontrolled ripening or softening of fruits or vegetables. It is possible that farmers in developing countries could benefit considerably from crops with delayed ripening or softening as this may allow them much greater flexibility in distribution than they have at present.

7. Tolerance to environmental stresses

Crop plants are subjected to variety of stress conditions like leaf drought, high salt content of soil, excessive moisture etc. Identification and isolation of the most important gene involved in protection against stress conditions could help in development of transgenic plants having resistance/tolerance to different stress conditions.

A salt tolerance gene from mangroves (*Avicennia marina*) has been identified, cloned and transferred to other plants. The transgene plants were found to be tolerant to higher concentration of salt. The gut D gene from *Escherichia coli* has also been used to generate salt-tolerant transgenic maize plants. Such genes are a potential source for developing cropping system for marginalised land.

8. Pharmaceuticals vaccines from Transgenic Plants

The genetic engineering of plants to convert them into highly productive, relatively cheap and easy to handle co-reactors is one of the prime aim of many companies and institutions in the world-wide. Riptides of pharmaceutical interest have already produced in transgenic plants. The production of high molecular weight proteins in plants is also feasible. Human serum albumin has been synthesized in transgenic potato and tobacco, monoclonal antibodies have been produced in tobacco and many other proteins including antigenic proteins for vaccine production are now being produced in plants.

The scope for example was that plants could be used to produce plastics, such as polyesters now synthesized from non-renewable sources such as petroleum products. Early results look promising. The work made use of the fact that many bacteria synthesize and store natural biodegradable polyesters, such as poly-3 hydroxybutyrate (PHB). Only three genes are required to make PHB, and therefore when transformed into plant *Arabidopsis* the transformed plants made some PHB. All these manipulations may turn out to be starting points for the production of environmentally safe biodegradable polymers. Plants are also the source of large variety of biochemicals which are metabolites of both primary and secondary metabolism (10). These include alkaloids, quinons, tarpenoids steroids, polyphenols, pigments, perfumes etc. It is believed that less than 10% of medicinal plants have been identified and characterized and the potential exist to use genetic engineering in a way that increases yields of these medicinal substances once identified.

Researchers are currently investigating the potential for GM technology to produce vaccines and pharmaceuticals in plants. This could allow easier access, cheaper production and alternate way to generate income. Vaccine against infectious diseases of the gastrointestinal tract have been produced in plants such as potato and bananas. Another appropriate target would be cereal grains. An anticancer antibody has recently been expressed in rice and wheat seeds that recognize cells of lung, breast and colon cancer and hence could be useful in both diagnosis and therapy in future (11). Therefore, the development of transgenic plants to produce therapeutic agents has immense potential to help in solving problems of disease in developing countries.

Biotechnology will have a major impact on agriculture in coming decades. The synergistic interplays of *in vitro* culture and genetic engineering techniques have made it actually or potentially possible to transfer any natural or *in vitro* modified gene to many plant species. A whole spectrum of gene technology is now routinely and successfully applied to a wide range of problem in plant biology, pathology, breeding and plant improvement in general. Of the various techniques used, somatic embryogenesis, double haploids, monoclonal antibody based diagnostic and *Agrobacterium* mediated and biolistics based gene transfers are most important. Transgenic plants in addition to enhancing crop productivity, provide a means to understand the mechanism of gene expression such as tissue specific expression of light regulated genes, male sterility and seed storage protein genes etc.

Further, fundamental studies in plant molecular biology and biochemistry are necessary for further biotechnological advancements in agriculture. Unless characterization of key metabolic processes is carried out, it will not be easy pyramiding genes for hyper expression of characters contributing to increased disease resistance, herbicide and stress tolerance. The identification and characterization of useful genes is still in its infancy and so is the work on regulation and desired levels of expressions of genes of economic importance.

The development of tissue specific and inducible promoter system is also a priority so that foreign gene expression can be in the appropriate tissue or can be 'switched on' by specific biochemical or physical treatments. The tremendous commercial potential of genetic engineering of plants will depend upon the successful transfer of a desired gene such that either higher expression of the gene is achieved or expression is directed tissue or stage specific of crop plant. Biotechnology for developing countries of course is surely beneficial alternative if appropriate technologies are employed for specific and need-based problems.

CHAPTER

9

RECENT ADVANCES IN PLANT GENETIC TRANSFORMATION

E. Sivamani

Plant transformation technology has become a versatile platform for cultivar improvement as well as for studying gene function in plants. This success represents the culmination of many years of effort in improvement of tissue culture techniques, transformation technology and advancement in molecular genetics.

Plant transformation through genetic engineering has opened new avenues to modify crops and provide new solutions to address specific problems in agriculture. The powerful combination of genetic engineering and conventional breeding programmes permits useful traits encoded by transgenes to be introduced to commercial crops within an economically viable time frame. There is great potential of genetic transformation of crops to enhance productivity through increasing resistance to diseases, pests, environmental stress and by qualitatively changing the seed composition. Plant 'factories' are being designed for high volume production of pharmaceuticals, nutraceuticals and other beneficial chemicals (Hansen and Wright, 1999).

Principal Investigator, Plant Transformation Laboratory, MAHYCO-Life Sciences Research Centre, PO Box 76, Jalna, Maharashtra-431 203, India

Transgenic plants are becoming drug delivery devices, with vaccines for HIV and rabies being synthesized in them (Yushibov *et al.,* 1997). Bananas have been engineered to produce edible vaccines (May *et al.,* 1995). With the establishment and expansion of genomics programme, a much broader range of genes with potential from crop improvement are being identified and in some cases, tailored and/or redesigned for further enhancement of their properties within specific crops (Estruch *et al.,* 1997). This has further intensified the interest in developing efficient plant transformation technologies to be able to concurrently test and capture the value of these genes.

Advances in tissue culture, combined with improvements in transformation technology, have resulted in increased transformation efficiencies. In recent years, many crops, previously classified as recalcitrant because they were not responsive to the overtures of genetic manipulations, have now been transformed. In many instances, the technology has been pushed even further to introduce the gene of interest directly into elite cultivars. The best example for this is rice and wheat. Elite varieties of cereals such as rice and whet are being transformed with much an ease and with a comfortable efficiency (Tang *et al.,* 1999; Sivamani *et al.,* 2000). As much as 14 transgenes have been introduced and expressed in rice (Chen *et al.,* 1998).

Tissue Culture

A tissue culture stage is required in most current transformation protocols to ultimately recover plants. Plants are regenerated from cell culture via two methods, somatic embryogenesis and organogenesis. Both are controlled by plant hormones and other factors. added to the culture medium. Plants such as soybean (Trick *et al.,* 1997), banana (May *et al.,* 1995; Sagi *et al.,* 1995) and sugarbeet (Krens and Jamar, 1989; Hall *et at.,* 1996) can be regenerated via either method so the choice is dependant upon which gives the best yield or the easiest outcome.

Somatic embryogenesis is the generation of embryos form somatic tissues such as embroys, microspores or leaves.

Proliferating somatic embryos in liquid culture or on solid medium are suitable targets for transformation because the origin of proliferating embryogenic tissues is at or near the older embryos and thus readily accessible to DNA delivery. Embryogenic tissues, in general, are very prolific, homogenous and normally allow recovery of non-chimeric transformants, because of the assumed single cell origin of somatic embryos (Maheshwari *et al.,* 1995). The tissue culture approach in monocots is generally chosen because callus is easily initiated from the scutellum of embryos from immature or mature seeds after exposure to auxins. The concentration and choice of auxin is to some extent dependent on the genotype and species.

Maize is reported to produce two types of somatic embryos (Tomes and Smith 1985): type I somatic embryos occur as group of fused embryos and type II occur as discrete single embryos in a cluster. Type I callus is convenient for biolistic transformation (Koziel *et al.,* 1993) and type II callus is preferentially chosen for protoplast isolation (Shillito *et al.,* 1994). Depending on the transformation experiment it is thus possible to initiate the callus type through tissue culture manipulation.

The use of embryogenic tissue can have some limitations. It can be labour intensive to establish and maintain the culture and recovery of plants can be a long process, with the risk of encountering morphological abnormalities and sterility. This system also requires a constant source of material to initiate new embryogenic cultures. This inconvenience has prompted the use of mature seed derived embryogenic calli as an alternate source (Sivamani *et al.,* 1996).

Transformation Systems

Many methods of transformation of plants have been reported. Among them are *Agrobacterium* mediated (Horsch *et al.,* 1985), protoplast transformation (Paszkowski *et al.,* 1984), electroporation (Fromm *et al.,* 1985), pollen tube pathway (Luo and Wu, 1988), Particle bombardment (1987), *in planta* transformation (Bechtold *et al.,* 1993; Clough and Bent, 1998). Successful and reproducible transformation of plants depends on, (1) target tissue competent for propagation or regeneration, (2) efficient DNA delivery method, (3) selection agents for transgenic tissues, (4) ability to recover fertile transgenic plants at a reasonable frequency and which is genotype independent, (5) a tight timeframe in culture to avoid somaclonal variation and possible sterility. At present the following three techniques appear to fulfil the above criteria. They are (1) Protoplast transformation, (2) Biolistic or microprojectile bombardment (3) *Agrobacterium* mediated transformation.

Protoplast Transformation

Of the three systems mentioned above, the protoplast method is the most cumbersome one and involves many steps. Protoplasts are isolated either by mechanical or by an enzymatic process to remove the cell wall. This results in the production of a suspension containing millions of individual cells and therefore offers the advantage of probable single cell targets. Protoplasts are frequently obtained from an established suspension cell line of callus initiated from imature embryos, immature inflorescence, mesocotyls, immature leafbases, and anthers (Maheshwari *et al.*, 1995). Technological advancement in cryopreservation, has led to store the suspension cells from plants for a long time without changing their regenerability etc. (DiMaiyo and Shillito, 1989). Protoplasts can either be transformed by Agrobacterium or by direct DNA uptake methods facilitated by polyethylene glycol, electroporation etc.

In some species, protoplasts can be isolated from leaf mesophyll This technology was fine turned to sophistication when stomatal guard cells from sugar beet leaf protoplasts were isolated and identified as the cells of choice for transformation because of their totipotence (Hall *et al.*, 1996).

Biolistic Transformation

Through this method the microprojectiles, usually tungsten or gold coated with DNA, are propelled into target tissue by acceleration. The source of acceleration used to be gunpowder (Klein *et al.*, 1987) or gases like Helium or by electrical discharge (McCabe and Christou, 1993). This method can introduce DNA into virtually any tissue from any cultivar-success depends critically upon the ability of the target tissues to proliferate and given rise to a fertile plant.

Minor modifications in the standard protocol have yielded tremendous improvements. These include: preculture of the plant material, use of baffling screens (Vain *et al.*, 1993a), use of small size microprojectiles (Randolph-Anderson *et al.*, 1995), osmotic pretreatment of the target tissue in a medium containing an osmotic agent (Vain *et al.*, 1993b). Randolph Anderson *et al.*, (1995) reported a dramatic increase in the transformation efficiency of an elite maize line under these conditions.

Some of the pitfalls of this technique include complex pattern of integration of transgenes in the genome, difficulty in delivering long fragments of DNA etc. Some of these problems can be circumvented by using linear transgene constructs (Fu *et al.*, 2000).

Agrobacterium-mediated Transformation

The natural ability of a particular genus of soil bacterium, *Agrobacterium sp.* to transform plants is exploited in this method. The *Agrobacterium* system is attractive because of the ease of the protocol coupled with minimal equipment costs. Moreover, transgenic plants obtained by this method often contain simple copy insertions. These advantages were a driving force to adapt this system to many different crops including monocots. High efficiency super binary vectors with extra copies of the vir genes or vectors with mutations that enhance the virulence gene expression have been constructed (Hansen and Chilton, 1999). With all these advancements it is now possible to transfer large fragments (150 kb) into plant nuclear genomes (Hamilton *et al.*, 1996). These days the efficiency of transformation in monocots using *Agrobacterium* has increased dramatically (Smith and Hood, 1995). Efficient and routine *Agrobactereium* mediated transformation of rice (Hiei *et al.*, 1994), wheat (Cheng *et al.*, 1997), barley (Tingay *et al.*, 1997), maize (Ishida 1996), sugarcane (Arencibia *et al.*, 1998) has been reported.

Alternative methods of transformation

There is a perpetual quest to find more efficient and economical methods for plant transformation, some of these are described here but are often not used because of their poor reproducibility and impracticability at present. Transgenic plants have been recovered after DNA delivery using silicone carbide whiskers (Thompson *et al.*, 1995). The electroporation method (Fromm *et al.*, 1985) is not often used because of its low reproducibility. Microinjection of DNA into zygotes led to recovery of transgenic plants (Leduc *et al.*, 1996). Uptake of naked DNA by plant can also be facilitated by laser beam (Hoffman, 1996). However, these options, which are still at the early stages of development, can be tedious and require sophisticated equipment.

In Planta Transformation

There is considerable interest in developing plant transformation methods that exclude tissue culture steps and rely on simple protocols. These methods are called in planta transformation because transgenes are generally delivered into intact plants in the form of naked DNA or from *Agrobacterium*. This concept is exemplified by the successful transformation method developed for *Arabiadopsis*. In this method *Arabidopsis*, flowers are infiltrated of dipped into an *Agrobacterium* suspension and subsequently some of the harvested seeds are transgenic (Bechtold *et al.*, 1993; Clough and Bent 1998). Although the overall efficiency is low, the sheer number of seeds recovered for screening and the ease of the method makes it an extremely attractive alternative.

A similar approach utilizing naked DNA has been attempted in other plants. Cotton transformants were recovered following injection of DNA into the axil placenta about a day after self-pollination (Zhou *et al.*, 1983). Similarly a mixture of DNA and pollen was either applied to receptive stigmatic surfaces or DNA was injected into rice floral tillers (Langridge *et al.*, 1992). In another approach, meristematic cells located in the apical dome of nodal buds have also been considered ideal recipients for transgenes because growth and development occur in this area. Excised apical meristems from embryonic axes and shoot tips have also been used targets for transformation by particle bombardment or by *Agrobacterium* inoculation (Park *et al.*, 1996). Using a similar approach, Rohini and Sankara Rao (2000 developed a system for transforming peanut. Using the lateral cotyledonary meristems, Bean *et al.*, (1997) developed a system to transform pea. These procedures, intriguing as they are, are impractical at present because of their low reproducibility.

Plant transformation remains an art because of the unique culture conditions required for each crop species. Transformation technologies have advanced to the point of commercialization of transgenic crops. Technical improvements that have the greatest opportunities for new approaches are probably in the area of *in planta* transformation. With further increase in the efficiency of transformation in this method, and extension of this technique to elite commercial germplasm, production costs can considerably be reduced which will ultimately lead to lower cost to the consumer.

Status on Applications of Transgenic Technology

Using the technology discussed above, not only the ability to genetically transform a wide variety of crops species has been enhanced, but also the capacity to generate variability for a range of economically important traits in crop plants has been established. Transgenic crops are under commercial cultivation mainly in the developed countries. In developing countries like India, several transgenic crop lines are at different stages of development or field-testing (Rai and Prasanna, 2000). It has been reported that more than 100 plant species are growing in laboratories, greenhouses or in the fields for experimental purposes (Persley, 2000). Transgenic crop with improve agronomic traits such as herbicide tolerance, pest and disease resistance, delayed ripening, pollination control has already reached consumers.

During 1986-99, nearly 45 countries have conducted more than 25,000 transgenic crop field trials on about 60 crops and over 70 genetically modified crops have been registered for commercial cultivation in more than 10 different countries

(Rai and Prasanna, 2000). The area under cultivation of transgenic crops has been steadily expanding, in 1996 it was 1.7 million hectares, 11 million hectares in 1997, 27.8 million hectares in 1998 and 39.9 million hectares in 1999 (James C, 1999).

Currently 10 prominent traits including herbicide tolerance, insect resistance, product quality and virus resistance, are particularly targeted through genetic engineering for development of transgenic crops. More than 16 crops that are engineered with the Bt endotoxin gene for insect control are already commercialized. Using coat protein mediated virus resistance technology more than 20 plant species have been field tested for virus diseases (James, 1998).

In a short span of 5 years (1995-99) revenues from transgenic crop have increased approximately 30 fold; from 75 million US$ in 1996 to 2.3 billion US$ in 1999 (Serageldin, 2000). Further, the analysts predict that global market for transgenic crops would reach 3 billion US$ in 2000, 8 billion US$ in 2005 and 25 billion US$ in 2010 (Serageldin, 2000).

With the repidly growing population and shrinking agricultural resorce-base, India has realized the importance of biotechnological tools to implove its food, feed and fiber production. Several Institutes under ICAR and CSIR besides other reputed research organizations and Universities are actively involved in research on transgenic crops such as rice, cotton, potato, brinjal, tomato, cabbage, cauliflower, mustard, tobacco etc. Some of these researches are in their final field testing stages and the products are expected in the market in the next 2-3 years.

References

1. Arencibia *et al.,* 1998. Transgenic Research 7: 213-222.
2. Bean *et al.,* 1997. Plant Cell Reports 16: 513-519.
3. Bechtold *et al.,* 1993. CRC academic science, Paris. Life Sciences, 316, 1194-1199.
4. Chen *et al.,* 1998. Nature Biotechnology 16(11): 1060-1064.
5. Clough and Bent, 1998. Plant Journal 16: 735-743.

6. Dimaiyo and Shillito, 1989. Journal of tissue culture methods 12: 163-169.
7. Estruch *et al.*, 1997. Nature Biotechnology 15: 137-141.
8. Fromm *et al.*, 1985. PNAS 82: 5824-5828.
9. Fu *et al.*, 2000. Transgenic Research 9: 11-19.
10. Hall *et al.*, Nature Biotechnology, 14: 1133-1138.
11. Hamilton *et al.*, 1996. PNAS 93: 9975-9979.
12. Hansen and Chilton, 1999. Current Topics in Microbiology and Immunology, Vol. 240. Plant Biotechnology: New Products and Application, Springer-Verlag.
13. Hasen and Wright, 1999. Trends in Plant Science 4(6): 226-231.
14. Hiei Y *et al.*, 1994. Plant Journal 6: 271-282.
15. Hoffman, 1996. Plant Science, 113: 1-11.
16. Horsch *et al.*, 1985. Science 227: 1229-1231.
17. James C, 1999. ISAAA Brief, Ithaca, NY, USA.
18. James C, 1998. APSA publication.
19. Klein TM, Wolf ED and Sanford JC. 1987. Nature 327: 70-73.
20. Koziel *et al.*, 1993. Biotechnology 11: 194-200.
21. Krens and Jamar, 1989. Journal of Plant Physiology 134: 651-655.
22. Langridge *et al.*, 1992. Plant Journal 2:631-638.
23. Leduc *et al.*, 1996. In Vitro Development Biology 10:190-203.
24. Luo and Wu, 1988. Plant Molecular Biology Reporter 6: 165-174.
25. Maheshwari *et al.*, 1995. Critical Reviews in Plant Science 14: 149-178.
26. May *et al.*, 1995. Biotechnology 13: 486-192.
27. McCAbe D and Christou P. 1993. Plant Cell, Tissue and Organ Culture 33: 227-236.
28. Park *et al.*, 1984. Plant Molecular Biology 32: 1135-1148.
29. Paszkowski *et al.*, 1984. EMBO Journal 3:2717-2722.
30. Prsley GJ, 2000. In: proceedings of International Conference on Agricultural Biotechnology and the Poor. Washington DC., 21-22 Oct 1999. Consultative Group of International Agricultural Research, Washington DC, USA, pp. 3-21.

31. Rai. M and Prasanna BM. 2000. Transgenics in agriculture. IARI, New Delhi. pp. 144.

32. Randolph-Anderson *et al.*, 1995. BioRad Bulletin 2015.

33. Rohini and Sankara Rao. 2000. Plant Science 150:41-49.

34. Sagi *et al.*, 1995. Biotechnology 13:481-485.

35. Serageldin, 2000. In: proceedings of International Conference on Agricultural Biotechnology and the Poor. Washington DC., 21-22 Oct. 1999. Consultative group of International Agricultrual Research, Washington DC, USA, p. 25-31.

36. Shillito *et al.*, 1994. The Maize handbook, Pringer pp. 695-700.

37. Sivamani *et al.*, 1996. Plant Cell Reports 15: 322-327.

38. Sivamani *et al.*, 2000. Plant Science 155: 1-9.

39. Tang *et al.*, 1999. Planta 208: 552-563.

40. Thompson *et al.*, 1995. Euphytica 85: (1-3) 75-80.

41. Trick *et al.*, 1997. Plant Tissue Culture Biotechnology 3:9-26.

42. Tomes and Smith 1995. Theoretical and Applied Genetics 70: 505-509.

43. Vain *et al.*, 1993a. Plant Cell Tissue and Organ Culture 33:237-246.

44. Vain *et al.*, 1993b. Plant Cell Reports 12:84-88.

45. Yushibow *et al.*, 1997. PNAS 94:5784-5788.

46. Zhou *et al.*, 1983. Methods in Enzymology 101:433-481.

CHAPTER 10

IMMOBILIZED ENZYME AND MICROBIAL BIOTECHNOLOGY IN BIOPROCESSING, FERMENTATION, BIOREMEDIATION, AGRICULTURE AND BIOSENSORS

S.F.D'Souza

Biotechnology is currently considered as an useful alternative to conventional chemical process technology. A developing country like India, with its increasing manpower and increasing pressure on natural resources; food production, processing and utilization of renewable resources are important for sustainable economic growth and development. From a technological point of view, enzyme and microbial technology seems to be best suited and most relevant to meet this requirement. Enzyme and microbial technology has influenced the process industry significantly in the recent years by improvement of existing processes towards achieving a better quality product at reduced cost, as well as in the production of unique products. The current demand for better utilization of renewable resources and pressure on industry to operate within environmentally compatible limits has also been a stimulus to the development of new eco-friendly industrial bioprocesses. One of the techniques which has played a significant role, is the immobilization of enzymes and

Nuclear Agriculture and Biotechnology Division, Bhabha Atomic Research Centre, Trombay, Mumbai-440 085, India.
E-mail: sfdsouza@apsara.barc.ernet.in

cells for their economic reuse under stabilized conditions in continuous bioprocesses. Immobilized enzyme and microbial technology will create newer avenues in bioprocessing, fermentation, bioremediation, agriculture and biosensor fields.

Enzymes as bioprocessing aids

Major applications of enzymes have been in food processing specially in the starch, juice and wine, brewing, distillery, banking and dairy industries. The future applications may be in the production of high value food additives like flavour compounds, sweeteners, antioxidants and nutritional supplements. Other applications are in the pharmaceutical, textile and detergent industries. In the future, enzymes may revolutionize the leather and paper processing industry by development of environmentally less polluting technologies. In addition to enzymes like proteases, amylases, pectinases, cellulose, lipases, oxidases, isomerases, xylanases, phytase etc., there is a great interest in finding novel enzymes with unique stability and specificity characteristics. Even though thermostable enzymes have been often preferred there is a current interest in cold active enzymes. These enzymes exhibit very high activities even at low temperatures and are gaining importance especially in food, pharmaceuticals and detergent industries.

Biotechnology has influenced enzyme industry significantly in the recent years especially in the more efficient production of enzymes, their stabilization and their economic reuse. Biodiversity is widespread among natural populations of microbes. Selective screening methods for discovering the rate genera and species, application of the knowledge on microbial biodiversity to discover novel and industrially useful enzymes and the need for an integrated approach to combine classical microbiology with developments in modern biotechnology for sustained advances in research and development are essential in the future. Molecular genetics and recombinant DNA techniques are being used to create strains that can over produce the desired enzymes and also in the production of non-microbial enzymes like calf chymosin. Protein engineering, either through site directed mutagenesis or through chemical modification can help in designing enzymes to work under varied environmental conditions of temperature, pH and salinity. In the future computer aided protein engineering will help in making a rationale design of proteins with new properties possible. Another important development is 'abzymes', the catalytic antibodies; it may be possible to produce 'smart' enzymes which catalyze reactions for which no known enzyme exists in nature. It will also be possible in the future to continuously produce enzymes using transgenic plants grown hydroponically.

Potentials of enzymes to function in non-aqueous environment has opened up newer routes for the biosynthesis of a variety of compounds. It is now well established that hydrolytic enzymes like lipases and proteases can catalyze the reverse reactions like synthesis of a peptide or an ester bond in an organic medium. Enzymes in organic solvents also help in the streo and regiospecific (trans) esterification reactions and synthesis of optically active pure compounds. Techniques are now available to obtain cross-linked enzyme crystals as well as organic-solvent-soluble enzymes for use in organic solvents.

The degree of purity of commercial enzymes ranges from crude preparations to highly purified forms depending on their applications. Enzymes that are employed in therapeutics or diagnostic/analysis or for highly selective biotransformations must often be prepared in high purity form. Developments in more efficient downstream processing systems like cell disruption techniques, membrane filtration, aqueous two phase, reverse micelles, biospecific affinity techniques can play an important role in the extraction, purification and concentration of enzymes. rDNA technology may help in the future in tailoring enzymes (fusion proteins) for their easy purification using biospecific affinity techniques. Enzymes that are used, as commodity products must be prepared cheaply and often may not need extensive purification. Past experience clearly suggests that we must learn to work with crude enzyme preparations for economic reasons. In this direction selection of microbial strains which produce minimal contaminating enzymes (like cellulase free xylanases) would be preferred in the future to minimize downstream processing costs. The other important development has been the use of permeabilised cells as an economical source of intracellular enzymes. Number of techniques have been developed at BARC for the permeabilisation of cells.

The major limitation in the economical exploitation of enzymes in industry is their instability and solubility in aqueous media thus restricting their one time use in batch processes. Immobilizing them on to insoluble polymeric supports can help in retaining them in a proper reactor geometry for their economic reuse under stabilized conditions. Immobilized biocatalysts offer several other advantages, notable among them is the availability of the product, in greater purity. Basically it is now possible to use very crude enzyme preparations or permeabilised cells or cell lysates in an immobilized form without contaminating the final product. A number of immobilized enzyme and nonviable cell based processes have become industrially feasible. Some of these include production of 6-amino penicillanic acid (penicillin acylase), resolution of DL amino acids (amino acid acylase), invert sugar (invertase) high fructose syrups (glucose isomerase), lactose hydrolysed milk (lactase) etc. This has been a major research activity in the author's laboratory and a number of new techniques and supports have been developed for the immobilization of enzymes, cells and enzyme-cell conjugates.

Fermentation

In fermentation technology the key areas are in the production of industrial enzymes, antibiotics, vitamins, biofuels, biosurfactants etc. In addition to obtaining novel strains one of the important challenges for the biotechnologists in the future is in improving the fermentation techniques. The classical fermentation suffers from various constraints such as low cell density, nutritional limitations and batch mode of operation with high down times. It has been well recognised that microbial cell density is of prime importance to attain higher volumetric productivity. The major limitation in the development of continuous fermentation process has been the wash out of cells from the bioreactor. Use of flocculating strains, cell recycle and membrane reactors are being investigated to solve some of these problems. The immobilized viable cell technology can eliminate most of the constraints faced with the free cell systems. The remarkable advantage is the freedom to determine the cell density prior to fermentation. It also facilitates operation of fermentation on a continuous mode without cell wash out even at high dilution rates. The immobilized cell technology process also decouples microbial growth from cellular synthesis of favoured compounds. Immobilized viable cells have been extensively studied in the continuous production of fuel ethanol, amino acids, organic acids, enzymes etc. Studies from our Lab have shown the application of this approach in the rapid removal of sugars through fermentation from perishable food commodities like milk and eggs. In the case of enzyme production using microbes, solid state fermentation holds tremendous potential in view of the high volumetric productivity, less effluent generation and requirements of simple fermentation equipments. In addition to microbial cells the fermentation technology using bioreactors is also gaining importance for the production of high value compounds using plant and animal cell cultures. Immobilization of such cells have been shown to offer them stability against shear force when used in continuous stirred tank bioreactors. Other future developments will be in integrating downstream processing with bioprocessing/fermentation and also in the development of appropriate bioreactors.

Bioremediation

Immobilized cells have a number of advantages in the development of waste bioremediation processes, as it solves the problems associated with solid-liquid separation in setting tanks. Biological wastewater treatment, quite different from controlled fermentation, has unique problems associated with fluctuation of influent quantity and constituents of toxic chemicals. Also many such processes exhibit substrate inhibition kinetics. One of the important advantages of immobilization has been in the protection of cells from the external environmental perturbations by the toxic compounds in the effluents. In addition to organic waste, bioremediation is also gaining importance as an alternative technology for the management of heavy mental and radionuclide waste. Microbes can be used to concentrate such metals from dilute waste steams either through biosorption or bioaccumulation processes. Unlike the organic toxicants which are biodegradable heavy metals are immutable at an elemental level. Any successful bioremediation process necessitates the retention of biosorbent in a reactor geometry. Immobilization/pelletisation of biomass has been found to be an essential component of this process. Transgenic phytoremediation has been found to be useful in the cleaning up of contaminated soils. In this respect more research is required in the future, in understanding the plant and rhizosphere microbe interactions. Work at BARC has concentrated in the preparation of biobeads or bioresins for the treatment of organic, heavy metal and radionuclide waste.

Agricultural applications

Immobilized cells are being investigated as an alternative technology for a variety of other environmental applications in agriculture, biocontrol, pesticidce application and pollutant (eg pesticide) degradation in contaminated soils. One of the major limitations for the applications of microbial consortium into soil is their survival under biotic and abiotic stresses like drought (wet-dry cycles), salinity, phage attack, heavy metal toxicity etc. Immobilization of the microbes through encapsulation or entrapment in certain defined polymers has been shown to afford protection to cells under such adverse conditions. Immobilization can help in devising newer strategies in the future in the introduction of microbes into the soils. Using this approach it is now possible to provide the microbial consortium with more defined microenvironment, quite different from that encountered directly otherwise in soil environments. In this context encapsulation process adds a modicum of control, potentially becoming a miniature reactor in the environment. Other advantage is the increased plasmid stability observed in the case of immobilized viable cells in view of their controlled growth. Immobilized cells can also act as synthetic inoculation carriers for the slow release of plant growth related organisms like *Rhizobium* into soils. In general entrapment technique is being increasingly investigated for the preservation of cultures as well as a source of continuous inoculum for a variety of fermentation applications. Immobilization can provide additional benefits for commercial purposes especially in terms of ease of storage and transportation and also in terms of biosafety features that limit contamination and bioaerosol formation. More research is required in the future, to establish the potential effectiveness of immobilization technology for use in the environment in varied soil systems. Some work has been initiated in this area at BARC especially in developing encapsulation materials with protective and magnetic properties.

Biosensors

To meet the present and emerging requirements of productivity, quality and reliability new analytical devices are being developed for quick and specific analysis. One of the rapidly emerging areas in the field of analysis is the biosensors which exploits the highly specific molecular recognition capability of certain biological materials. A biosensor consists of a biological sensing element such as enzymes, antibodies, DNA, receptors, organelles and microorganisms as well as animal and plant cells or tissues in close contact with a transducer such as an ion specific electrode, thermistor, optical fiber, piezoelectric crystal, silicone water, conducting polymer etc. The biological signal arising due to its interaction with the analyte can result in an electronic or allied signal that can be easily measured, amplified and documented. Technique for the immobilization of the biomaterials has played a significant role in biosensor field. Immobilization not only brings about the intimate contact of the biological catalysts with the transducer, but also helps in the stabilization of the biological system thus enhancing its operational and storage stability. Biosensors generally provide a rapid and convenient alternative to conventional methods for monitoring chemical substances in fields as diverse as medicine, environment, fermentation and food processing. In medicine, biosensors can be used for the monitoring of blood glucose, urea, cholesterol, lactate, neuro transmitters etc. They are also gaining importance in fermentation for the monitoring of assimilable nutrients and intermediates and end products of fermentation and in food processing for shelf life assessment (freshness) of fish, meat and vegetables, microbial contamination and olfaction. Biosensors will also play a major role in environment monitoring like, BOD estimation, assay of toxic and mutagenic environmental chemicals and explosives like TNT and RDX. Genetically engineered microbes may also gain importance in the future as bioreporters. When such microbes metabolize the organic pollutants, the genetic control mechanism also turns on the synthesis of the enzymes like luciferase, which produces light in the presence of oxygen.

As the biosensor technology starts moving from the proof-of-concept stage to field testing under realistic process or waste monitoring conditions, the need for availability of stable biological materials will be important. Advances in biochemistry, molecular biology and immunochemistry in the future will lead to a rapid expansion in the range of biological recognition elements leading to more wider applications in the use of biosensors as alternative or newer analytical tools. Biosensors have been miniaturized extensively in the recent years, for this purpose it is necessary to develop suitable methods in the future for the microimmobilization of the biomolecules. Major advantages of biosensors will include simplicity of operation, low operating cost and could provide faster real time analytical data making it a cost effective relevant to developing countries. Few novel techniques have been developed in our laboratory for obtaining enzyme and microbial films either through surface adhesion or entrapment for use in biosensor applications.

With increasing emphasis on biotechnology, enzyme and microbial technology is acquiring a new dimension and is creating opportunities for expansion using multi-disciplinary approaches. Considerable progress has been made in the field of enzyme and microbial technology at the National level. A recent issue of Current Science (Vol. 77, No. 1, 1999) has devoted a special section on 'Fermentation-Science and Technology' highlighting the National scenario. The Presidential address delivered by Dr. (Mrs.) Manju Sharma,

Secretary, DBT at the 86th Session of the Indian Science Congress, 1999, which has appeared in a recent issue of Proceedings of the National Academy of Sciences, India (Vol. 69, Part II, 1999) deals with an overview on 'New Biosciences: Opportunities and challenges as we move into the next millenium' with special reference to our National scenario. A few relevant reviews from the author have been indicated below.

CHAPTER

11

BIOTECHNOLOGICAL SUSTAINABLE DEVELOPMENT—AN INDIGENOUS WAVE OF INNOVATION

Sudhir U. Meshram

A major challenge in reference to environmental planning and production is to provide a helping hand to the economy in India and to meet healthy foodgrain needs of its over growing population. In developing countries like ours, adoption of any new technology in production sector is always hampered by a lack of necessary inputs and poor monetary capacity. Besides, with the advent of modern technological revolution, all kinds of impurities have began to be added to the natural ecosystem of air, water, as well as soil, causing irrepairable and irreversible damage to environment. Diverse range of pollutants in the form of gases, agricultural recalcitrant chemicals, particulates, oil spillage, solid waste etc. are effecting directly or indirectly even the most benevolent organisms. The pollution has indeed acquired the distressing dimension for the present as well as the future generations. Therefore, the restoration of environmental equality of the ecosystem has become our immense concern. This can be achieved by the application of microorganisms in most of the

---

Professor and Head, P.G. Department of Microbiology, L.I.T. Campus, Nagpur University, Nagpur-10 and Director, Rajiv Gandhi Vikas Biotechnology Centre, L.I.T. Campus, Nagpur University, Nagpur-440 010 (M.S.) India.

major production sectors which are responsible for the deteriorating environmental biospheric quality.

Application of microbial biotechnology is expected to go a long way in helping to restore the environmental quality of barren/wasteland too, and some measure of success has been achieved in this field by the author. On account of disturbed ecology, disturbed soil conditions and climate extremes, these sites have been rendered practically devoid of organic matter and are poor in supply of essential plant nutrients, and thereby rendering them inhospitable for plant growth. Therefore, development of such barren/wastelands also needs a biotechnological package which can improve such sites to bring back the ecological balance as an alternative measure for arable production units. The major arable crop production unit is already on the verge of extinction. With the use of proper soil amendments, the barren/wasteland sites can be developed to become amenable for producing the biomass as source of energy, and also to some extend of increasing production unit of the country so as enormous growth of population.

In view of escalating energy cost coupled with growing environmental pollution menace and financial constraints of small and marginal producers, it is felt necessary to adopt a strategy of 'Integrated approach' in crop production sector by using judicious combinations of chemical fertilizers, pesticides, organic manures and microbial biotechnology. The use of chemicals in the form pesticides are criticized owing to higher toxicity vis-à-vis the beneficial activity of biosphere ecosystem eliminating the natural enemies of pest than pest itself.

Broadly speaking, biotechnology includes bioconversion of organic wastes, application of biofertilizers, biopesticides, tissue culture, modification of organisms to obtain desired characteristics, etc.

The introduction of biofertilizer and biopesticides accrued enormous benefits especially in developing countries to yield increased production at affordable cost. The use of biotechnology by the author and his research group have proved to be useful for improving the fertility of even barren/ wasteland and also for creating pollution-free environment.

Biofertilizers enhance soil fertility and yield of crops by converting the unavailable sources of elemental nitrogen, bound phosphate and decomposed plant residues into available forms which help and facilitate the plants to absorb the nutrients. The microorganisms involved in nitrogenous biofertilizers programme i.e. *"Integrated Nutrient Management"* (INM) are symbiotically grown in plants as well as free 'N' fixing prokaryotes like *Blue Green Algae, Azotobacter, Azospirillum* etc. Most of these bacterial species are heterotrophs and thus require an exogenous source of carbon and energy. Blue Green algae and Azolla are self supporting systems being photoautotrophics in nature, form important component of biofertilizer industry especially in developing countries like India.

Human needs in terms of agricultural production change with increasing population. The pressure for increased food supply become so great that Microbiologists, Environmental Genetic engineers and plant breeders have to explore new approaches for a quantum jump in crop yields without disturbing the ecological balance by adopting *"Integrated Sustainable Development Programme"*.

According to the Annual Report of DBT (Ministry of Science and Technology, Government of India, 1989-90), crop losses due to insects and pathogens are very heavy in India. About 6000 crores of rupees worth of agricultural production is lost every year due to pests alone. Likewise, 200 tons of fish mortality is attributed in Vidarbha region alonge (Maharashtra State) due to infestation of fishes. In view of this, the newly acquired knowledge of antagonism phenomenon also known as 'Biological Control Method' is most useful in responding to this challenge. The biological control of pests is a scientific tool which explores secrets of life existing in the biosphere and at the same time provides us with useful means to cope up with many complex social problems, thus becoming an intersection which links environmental science with society.

Chemical pesticides are counterproductive in two major ways. First, they cause a resurgence or increase in the pest population by developing resistance of often used chemicals.

Second, they cause upsets in the ecological balance. In nature, more numbers of potential pests are kept in check by their natural enemies which are being eliminated due to application of recalciterant chemical pesticides. Ultimately several problems have cropped up due to intensive and indiscriminate use of these chemical pesticides. The insects that were considered minor pests have become serious pests when their natural enemies present in environment are destroyed by such chemical pesticides. This also involve substantial cost in terms of human health.

As compared to earlier pesticides, the modern pesticides are much more potent and difficult to handle. Every year, numerous cases of human and animal poisoning are reported. This is an alarming situation indeed. It is observed that some of the insecticides get accumulated in the soil and eventually enter biological life chain through plants or water source, upsetting the ecological balance. Such pesticides which were once highly appreciated, have now been banned recently in many countries in view of enormous environmental imbalances and discorders due to their persistence and biomagnification in environment of resistance among the insects, corrosiveness and harmful effects on non-target species.

In this backdroup, and attractive alternative involves the use of entomopathogens specific against insect species and antagonistic organisms capable of reducing the disease index or population of disease producing microbial flora. This process is known as biological control of pests by using biopesticides.

Biological control method is specific and generally does not disturb the natural ecosystem. They do not create blackash in terms of aggravating the problems that they are intended to remedy or by creating new problems as happens with chemical pesticides. It is also a low cost method as compared with the production, management and application of chemicals which concerns over the environmental consequences, the resulting legal constraints in regard to use of non-selective persistent pesticides and the development of resistance by target species. This has triggered off intensified efforts to formulate strategy of biological control agents for the suppression of these pestiferous insects, vectors and pathogens. The conceptualization and development of biological control has occurred within the framework of the gradual accumulation of biosphere life and ecological knowledge in the course of the progress of the human civilization.

The genetic engineering technology has also the potential to eliminate the serious environmental problems caused by synthetic pesticides. This has been achieved by moving genes into crop plants that code for synthesis of environmentally safe toxic compounds to protect them from insects and pathogens. The scientists have transferred a gene from *Bacillus thuringiensis* (the gene codes for delta endotoxin) into tobacco and tomato. The genetic engineers take it as a challenge to develop such type of transgenic plants that kill insects as effectively as chemical insecticides. Such replacement of insecticides responsible for on-site and off-site pollution would also be welcomed by the environmental activists. However, an introduction of transgenic seeds need to be studied carefully in terms of socio-economic condition of farmers and profitability posed by the multi-national Cos., inheritance environmental balance and utmost attention towards loss of local indigenous germplasm during the adoption process of genetic engineering transgenic technology.

The studies carried out in India and abroad by the present author on *"biological control of plant pathogens"* without any modification of natural probiotics have shown remarkable results in field conditions too. Recently, the research project carried out on *"Biological control of cereal plant pathogens by Azotobacter"* (Sponsored by ICAR, New Delhi) has significantly contributed to the concept of biological control phenomenon.

The phenomenon of increasing the growth and yield of several agricultural crops in field conditions is observed owing to a multiple action of this microbial technology *Azotobacter chrococcum* in soil i.e., nitrogen fixation, suppression of pathogenic organisms by producing inhibitory metabolites, production of growth promoting substances, stimulating effect on other beneficial organisms, and improved uptake of soil, phosphate. The investigation suggests an improvement of seed germination with suppression of seed borne fungi of cereal crops with saving of 20-40 kg N per hectare.

Similarly, the application of *Azobacter chrococcum, Rhizobium* spp. and mycorrhizae in barren/wasteland through fast growing trees have improved the fertility status of barren/ wasteland. These microbial inocula also increase the hard coated seed germination of *Subahul, Eucalyptus, Sewan., Cassia and Prosopis* with vigorous healthy seedlings stock in nursery. This investigation was sponsored by the Department of Non-conventional Energy Sources (Biomass), Ministry of Energy, Government of India. The field trials conducted on wasteland/barren land of L.I.T. Premises of Nagpur University by fast growing trees especially the Subabul test trees, have shown remarkable improvement in barren/wasteland fertility status with maximum biomass production as a result of soil and nursery stock inoculation with *Azotobacter chrococcum, Rhizobium* spp. and mycorrhizae.

These findings reveal that selection of plant species suited to the physico-chemical composition of the barren/ wasteland along with their most appropriate, well adopted and useful microsymbionts and asymbionts adopted to stress condition of soils is one of the essential criteria in barren/ wasteland development programme. Soil or nursery stock inoculation with specialized isolates of microorganisms known to improve physico-chemical properties of these soils with the manipulation of microbial profile responsible for healthy nutrient cycling is one of the significant corrective measures for amelioration of barren/wastelands. This has been confirmed in the investigation carried out recently by growing Wheat, Soyabean, Sorghum, Tur crops in R and D Projects sponsored by DBT, Min. of Sci. and Tech., Govt, of India entitled "Programmes Supported for Biofertilizer Technology Development and Demonstration in Backward region of Vidarbha" at Rajiv Gandhi Vikas Biotechnology Centre. Nagpur.

The other group of microorganisms have also been studied in our laboratory to control harmful insects and fish pathogens. These antagonistic microbes have spelt out a good hope in biological control method specially *Bacillus thuringiensis* has superceded all other microorganisms controlling the larvae of houseflies and mosquitos, the environmental vectors responsible for transmitting the pathogens of dreaded diseases like malaria, yellow fever, dengue, encephalitis, filaria, typhoid, paratyphoid, cholera, amoebic dysentery, food poisoning etc. Mosquito borne diseases afflict both developing and developed countries. The WHO under its special programme for biological control of vectors has included all aspects to be studied on *Bacillus thuringiensis*. The studies designed and conducted in our laboratory in the course of investigation revealed *Bacillus thuringiensis* to be as safe for mankind as to the other non-target species.

As for fishes, the pathogen of the genus *Klebsiella* causes a serious abdominal syndrome in the major carps, namely Catla, Rohu and Mrigal. Likewise, genus *Pseudomonas* also comprises certain species which cause several forms of infectious dropsy of carps.

According to various reports of researchers and Government organisations and NGO units, fish losses due to pathogens which enters in aquaculture through contaminated feacal matter, and emission of various toxic chemicals, industrial effluents including the pesticides in fresh and marine water are very heavy. Besides the use of chemicals and antibiotics for controlling fish diseases creates an environmental pollution detrimental to the life of fishes.

It has now been observed in the course of the investigation carried in our laboratory and field ponds that, there is an antidote to infections diseases. If the antagonistic bacteria *Bacillus thuringiensis* is introduced along with the pathogen at a certain concentration, the aetiological agent is unable to cause infection. This is ascertained by introducing the pathogen along with the antagonistic bacteria through ingestion method with fish food substrates and injected method between two scales and then inserted at an acute angle, on the left side of the ventral fin. Thus, the loss of several tons of fishes can be prevented with the application of natural form of biopesticide with an inovatived LDt^{50} value i.e. new concept of evaluating antagonistic efficacy of life biopesticides against live pathogens. This may boost up the quality of acquatic environment along with increased profit.

In regard to production of Biopesticides and Biofertilizers, an indigenous technique has been developed at pilot scale by developing a simple indigenous fermentor unit using natural locally available inexpensive materials in vidarbha region of Maharashtra State, India.

In conclusion, it may be said that an introduction of microbial biotechnology would be a boon for most of the sectors like Agriculture, Medical, Environmental and Industrial. The interest in natural integrated control has been stimulated in the main, by the failures and disasters thrown up by our total reliance on the use of highly budgeted organic chemical pesticides. During the past 30 years mankind has been almost completely dependent on synthetic organic pesticides. Today, the very properties that motivated people to use these chemical initially and were thought to be so useful with long residual action and toxicity for a wide spectrum of pests, are now reckoned to have brought about serious environmental problems. The failures of these chemical pesticides are mainly due to their adverse effects on natural enemies compounded by increasing the development of resistance factor by pests/ vectors to pesticides. As a result the earlier powerful chemical weapons are slowly vanishing but suitable replacements to these chemicals are scarce.

In response to the need of the hour, an exposure of microbial biotechnology can be an advantageous factor relieving the pressure on chemical fertilizers and pesticides without impairing productivity and also providing the additional yield and monetary returns by improving the fertility status of land. Further, according to the present investigation biotechnology in a water or soil body does not imbalance the food chain nor creates any hazard to the life and growth of flora and fauna.

However, an application of live form of biopesticides needs to be target oriented with an ascertanation of specific LDt value. If not so, soon or later we shall be in same cage again for similar situation that has arisen due to application of indiscriminate use of chemical pesticides.

Thus, the strategy of application of Biotechnology in sustainable development form can stand the Indian situation of depleting agricultural produce and growing barren lands in excellent stead, without in any way polluting the environment and disturbing the ecosystem which is so precious to all of us. This promising avenue is being vigorously explored through R and D transfer technology by us with its demonstration at tribal and rural regions uplifting their socio-economical welfare conditions in the present context of modernization.

Indeed, a wave of local indigenous innovation needs to be boost-up by the Government decision making bodies. Various communities and regions are taking imaginative steps to tackle economic, social, safety, and environmental challenges posed by new patterns of development. This wave of community-based activity was recently described in the final report of the National Commission on Civic Renewal, chaired by William Bennett and former Senator Sam Nunn.

While growth is essential to our continued economic prosperity, the individuals and communities involved in these partnerships are beginning to evaluate the costs of current growth patterns. They are questioning the economic cost of abandoning infrastructure in the city, only to rebuild in the suburbs. They are questioning costs to our quality of life from ever increasing traffic congestion. In other words, people and communities are trying to distinguish types of growth that solve and prevent problems from those that cause problems. They want to promote sustainable growth—growth of jobs, wages, educational achievement, and time with family—but not growth of pollution, poverty, commute times, and crime. Those who make such distinctions are not "no growth" advocates, or even "slow growth" advocates. They want the jobs, tax revenues, and amenities that development can provide. But they want it without degrading their environment, without unduly raising their local taxes, and without diminishing their quality of life.

REFERENCES

1. Sudhir U. Meshram and S.T. Shende 1982. Response of Maize to *Azotobacter chroococcum*. Plant & Soil Vol. 69 (2), 265-273.
2. Sudhir U. Meshram and S.T. Shende 1982. Total Nitrogen uptake by Maize with Azotobacter inoculation. Plant & Soil Vol. 69 (2) 275-280.
3. Sudhir U. Meshram and G. Jager 1983. Antagonism between *Azotobacter chroococcum* isolates and Rhizoctonia solani. Netherlands Journal of Pathology Vol. 89, No. 2, 191-197.
4. Sudhir U. Meshram 1984. Suppressive effect of *Azotobacter chroococcum* on *Rhizoctonia solani* and infestation of Potato. Netherlands Journalof Plant Pathology Vol. 90, No. 2, 127-132.
5. Sudhir U. Meshram *et al.*, 1988. Cultivation of *Azotobacter* on residual liquid from leaf protein extract. J. Microbiology Biotechnology Vol. 3 (2), 75-78.
6. Sudhir U. Meshram and S.T. Shende, 1990. Response of Onion to *Azotobacter chroococum* inoculation. J. Mah. Agril. Uni. 15 (3), 365-366.
7. Dolly Wattal Dhar, D.P. Singh and Sudhir U. Meshram 1989. Relative effectiveness/efficiency of different strains of Rhizobium on morphological attributes of *Leucaena leucocephala*. Indian Journal of forestry Vol. 12 (4), 310-313.
8. Sudhir U. Meshram *et al.*, 1993. Response of Seed-borne pathogens of cereal crops to *Azotobacter chroococcum* strains. Journal of Biol. Control Vol. (2), 87-92.
9. Sudhir U. Meshram 1993. Control of environmental pollution and improvement of wasteland through Biotechnology. Global Environmental Education-Vision of 2001 (Proc. Book of Environmental World Congress), 199-204.
10. Sudhir U. Meshram, Sonali Joshi and Sunil Pande 1994. Application of Microbial Biotechnology for improvement of Westeland. Proc. Book of Environmental Biotechnology (Sponsored by DBT, Ministry of Science & Tech. Govt. of India) Published by Distributed Information Sub-Centre, NEERI, 97-123.

11. Sudhir U. Meshram *et al.*, 1994. Microbial Technology for raising seedling of fast growing trees. Indian J. of Forestry Vol. 17 (3), 243-248.

12. Sudhir U. Meshram 1995. Rural and Tribal development in present context of modernization, Souvenir of Indian Pharmaceutical Association, 21-24.

13. Sudhir U. Meshram, Sonali Joshi and Sunil Pande 1995. Microbial Biotechnology for biomass production in Waste/ barren land. Agropedology Vol. 5, 83-90.

14. Sudhir U. Meshram, S.A. Peshwe and S.N. Joshi 1996. Bio-Fertiliser Technology for total Biomass Production of *Cassia siamea* on Weste/Barren land. Biofertilizer Newsletter Vol. 4 (No. 1), 6-7.

15. Sudhir U. Meshram, *et al.*, 1997. Response of Biofertilizers Biomass production of *Eucalyptus camaldulensis*. Annals of Forestry (International Journal) Vol. 5 (No. 1), 43-49.

16. Sudhir U. Meshram, Sonali Joshi, Ramdas Kamdi and Swati Peshwe 1989. *In vitro* Interaction of Microbial Biopesticides with fish pathogens prevailing in Aquaculture Food Industry. Journal of Food Science and Technology Vol. 25 (No. 2), 177-178.

17. Sudhir U. Meshram *et. al.*, 1999. Bio-control of fish pathogens associated with polluted Aqua-culture. Int. J. of Poll. Res. Vol. 18 (No. 4), 369-371.

18. Sudhir U. Meshram 1999-2000. Application of Biopesticides for controlling fish pathogens. Aquaculture Feed and Health Pub. Biotech Cons. India Ltd., New Delhi, 153-158.

CHAPTER

12

FOOD SECURITY THROUGH FISHERIES AND AQUACULTURE BIOTECHNOLOGICAL APPROACHES

A.S. Ninawe

Aquaculture has emerged as one of the most promising industries in the world with considerable growth potential. As one of the fastest growing and highly productive systems, can supply a cheap source of protein and also help alleviate poverty and reduce malnutrition among the rural masses. The per capita fish consumption of India is less than 11 kg per year. Thus we need to increase our per capita fish consumption level by increasing fish production. Our present fish production is 5.35 million tonnes, out of which 2.97 million tones is from the marine sector and 2.38 million tonnes comes from the inland sector.

Inland fish culture

The introduction of scientific composite fish culture technology has revolutionized the aquaculture sector of the country, raising the freshwater pond productivity from a subsistence level of 500-600 kg/ha/yr to a national mean level of about 2 tonnes/ha/yr. The development of intensive culture technology with production levels of 10-15 tonnesh/ha/yr during recent years is another breakthrough, showing the potential for expansion in rural ponds and tanks.

Director, DBT, Min. of Sci. & Tech. (Government of India), CGO Complex, Block-2, New Delhi-110 003, India.

In culture practices, both indigenous and exotic carps, viz. catla, rohu, mrigal, siver carps, grass carps and common carps accounted for bulk production. These culture species fetch good price in the domestic market. Technologies are available for breeding and culture of air breathing *(Clarius batrachus, Heteropneustes fossilis)* and non-breathing catfishes *(Wallago attu, Mystus seenghala, Pangassius pangasius* and can be used for organised catfish culture in the country. The freshwater prawns, *Macrobrachium rosenbergii and M. malcolmsonii* are already receiving keen attention with regard to the establishment of systematic farming and harvesting systems.

The development of hatchery technologies for the catfish species such as magur *(Clarias batrachus)* and singhi *(Heteropneustes fossilis)* along with freshwater prawns, *Macrobrachium rosenbergii* and *M. malcolmsonii* has opened up possibilities for further expansion of grow out culture systems.

Marine/backishwater fish culture

The development of technologies for coastal mariculture resulted in the rapid growth of this sector. Important brackishwater shrimps have been bred under captivity and reared successfully. Hatchery and culture technology has been developed for *Penaeus indicus, P. semisulcatus* and *P. Monodon*. Several other species are available for culture. However, presently the enterprise is concentrating on two species namely. *P. monodon* and *P. indicus*.

Shrimp farming sector has faced some setback due to intensification and adoption of unscientific management practices This has resulted in a huge loss to the shrimp industries due to disease outbreak. Alternative options for sustainable growth of aquaculture have been looked into through diversification.

To sum up the culture technological developments, carp and shrimp culture technologies have been developed for extensive, semi-intensive and intensive farming practices to suit different levels of investment in the venture. Technology for hatchery management of freshwater prawn has been developed as a viable package. Pearl production using freshwater mussels is a developing technology for which technology has been refined for native species. Integration of carp culture with agriculture and livestock rearing has been successfully developed. This technology is being used for recycling municipal sewage also. Azolla biofertiliser, vermicompost and biogas slurry have been used in aquaculture as eco-friendly measures. Techniques have been developed for multiple breeding of carps. Cost effective indigenous hormone has been developed for successful induced fish breeding. Diversification efforts were made in brackishwater aquaculture for cultivating diverse species like mullets, miklish, pearlspot, seabass and crustanceans. Technology packages for culture of edible oysters, pearl oysters, mussels and clams have been developed and transferred to field.

Shrimp aquaculture is largely depends on wild caught spawners. Hence, year round production of seed from captive broodstock is a priority in aquaculture research. As shrimp seed is an important input in culture system, hatcheries should concentrate on producing quality seed. Thus, we need to develop a system that can produce high performance nauplii using pond reared disease free broodstock.

Biotechnological applications

Biotechnological research and development are moving at a very fast rate. Modern techniques are being used to create new strains of fish, development of vaccines and diagnostics, biological control measures in aquatic food production system. These tools will help to improve the yields. The potential areas of biotechnology in aquaculture include the production of monosex population, transgenic culturable carp and catfish, construction of genomic libraries; cryopreservaiton of gametes; development of cell lines for genetic and virological studies; improved feeds and development of natural products etc.

The development of diagnostic reagents and vaccines by molecular biological techniques for disease control, use of probiotics in feed and production of genetically altered aquatic species have been seriously taken into consideration.

Seaweed biotechnology

The large scale cultivation of micro algae and the practical use of its biomass as source of certain constituents of various industrial uses has grown rapidly in recent years. The commercial algae production is restricted to very few species producing high value food. Seaweed protoplast fusion and somatic hybridization technique can be used for producing new cultivable strains of high quality seaweeds.

Genetic engineering

Genetic manipulation can enhance productivity growth. Thus attention has also been drawn towards genetic improvement of fish stock through selection and genetic engineering process. The potential of transgenic fish lies in developing transgenic broodstock lines. Combining transgenic technology with traditional selective breeding programmes can produce superior strains of fish. We need to produce our own indigenous gene constructs.

Fish vaccines and diagnostics

It is like that in the coming years biotechnology will have a great impact on the production of vaccines that will enable aquaculturists to protect fish against many important diseases and enhance production. There is need to develop vaccines against various bacterial, viral and parastic diseases of commercial important culturable fish species. Investigations on immunological responses need to be encouraged.

Biofertilisation/bioconversion

Possibilities have been explored in utilisaiton of various agriculture and agro-industrial by-products for aquaculture purpose, development of microbes for improving the utilisation nutrients from agro-industrial by products, increasing nitrogen fixation in ponds with blue green algae like Azolla. Thus genetic manipulation of nitrogen fixing bacteria to improve their nitrogen fixing capacity would be an important area in improving fertilisaiton quality in fish ponds.

Post harvest processing

Large quantities of fish are discarded at sea because it is uneconomic to preserve them and due to lack of infrastructural facilities. Application of biotechnology in fish processing and product development and value addition holds a good promise. Post harvest technologies can be developed for various products like protease enzyme useful in seafood industry, fish silage, chitin and chitosan etc.

Research areas for focus

- Studies on hybridization in freshwater prawn *Macrobrachium rosenbergii* for production of genetically improved strains. Improvement of stocks through genetic selection and genetic engineering.
- Induced breeding and domestication of commercially important species.
- Establishment of cell lines for virus isolation, cultivation and study of pathogens.
- Seaweed protoplast fusion and *in vitro* culture techniques to produce high yielding economically important species.
- Development of diagnostics and vaccines in aquaculture and use of probiotics and immonostimulants.
- Pathogen free broodstock development.
- Indigenisation of feed technology and production.

CHAPTER 13

EMANATED RECOMMENDATIONS

The Post-Graduate Department of Microbiology of the Nagpur University (India) hosted the First international conference on "Global Sustainable Biotech Congress-2000 AD" in Nagpur, Maharashtra State during 27th Nov. to 1st Dec. 2000, as a part of its Silver Jubilee Celebrations. The Conference was inaugurated at 10.30 AM on 27th Nov. 2000 at the hands of Dr. D.P.S. Verma, Professor, Molecular Genetics Biotechnology Centre, Ohio State University, Columbus (USA).

The Key-note address was delivered by Prof. Dr. D.P.S. Verma, FRCS (USA). The focal theme of the address was "Biotechnology Revolution-From Microbes to Man". The scholarly lecture had an electrifying effect on the delegates and guests present and it provided the delightful take-off for the enlightening voyage into the realm of Biotechnology with all its manifestations and ramifications in the subsequent technical sessions.

One of the unique features of the 1st International Conference is the organisaiton of "lab to land" Technical Sessions at three Villages Lohari Sawanga (Nagpur Distt.), Talodhi Balapur (Chandrapur Distt.) & Kelzar (Wardha Distt.) on 29th and 30th Nov. 2000. These parallel technical sessions were conducted at three villages in the local language, translated on the theme of "Role of Sustainable

Biotechnology". About 2000 Villagers participated in the sessions. The tremendous excitement generated in the villagers showed the vibrant response of the rural folk for the biotechnological innovations meant for their progress, development and benefit particularly in backward region of Vidharbha (M.S.), India.

The Technical Sessions dealt with 13 status papers of many front-line areas of biotechnology research, which include:

- Biotechnology and the principles of efficacy, safety and precaution
- Biotechnological sustainable development—An indigenous wave of innovation
- PUF-Immobilized cyanobacterial biofertilizers for rice
- Plant hairy roots: Nature's gift to biotechnologists
- Recant advances in genetic transformation
- Plant biotechnology applications in agriculture
- Human & Veterinary Plant-based production of vaccines: edible vaccines/pharmaceutical vaccines from transgenic plants
- Use of aquatic plants/wetlands system in wastewater treatment
- Immobilized enzyme and microbial biotechnology in bioprocessing, fermentation, bioremediation, agriculture and biosensors
- Microorganisms and global sustainability of paint industry
- Application of PCR in detection and monitoring of fungal plant pathogens and human cytomegalovirus in urine etc.
- Biotechnological approaches for food security through fisheries and aquaculture

About 450 delegates and invitees from renowned world R and D organisations and educational institutions have participated in the conference which include appartado de correos Bilbao (Spain), University of Mauritus (Reduit), Ohio State University (USA), University of Westminster, London (UK), Bureau of Pharmaceutical Assessment, Ottawa (Canada) and organisations Institutions from India such as the National Bureau of Soil Survey and Land Use Planning, (NBSS & LUP), Bahbha Automic Research Centre (BARC), Central India Institute of Medical Sciences, Central Institute of Cotton Research (CICR), MAHYCO, Tamilnadu Agricultural University, Banaras Hindu University. Indian Agricultural Research Institute (IARI), Mumbai University, Government Institute of Science (M.S.) MPKV, PKV, NBPGR, North Maharashtra University, Sardar Patel University (Gujarat), SRTMU (M.S.), University of Delhi, University of Roorkee, Department of Pollution Studies (Karad), NEERI, Nagpur, Aurora P.G. College, Hyderabad (AP), BSBA University, Nagpur University, Amravati University, Maharashtra University of Health Sciences, HAU (Haryana), ITRC (Lucknow) etc. To mention the more noteworthy Eminently; Prof. D.P.S. Verma (USA), Prof. Carlos M. Romeo-Casabona (Spain) & others who embellished the proceedings of the conference.

About 74 Research papers through Poster session were presented on various applications of biotechnology related to Environment, Industry, Agriculture, Medical Sectors etc. which generated lot of interest and enthusiasm among delegates, young researchers and students. The major thrust areas included;

- Recombinant DNA applications and their implications
- Molecular Pharmacology & Gene Therapy
- Plant Tissue Culture & Biotechnology
- Microbial Fermentation & Biotechnology
- Synthetic & Transgenic Edible Vaccines
- Immunology, Toxicology & Animal Biotechnology
- Sustainable Management of Waste land/Barren land, Crop Productivity, Aquaculture and Bio-diversity

Another important feature in conference was the Scientific Model/Poster presentation competition among undergraduate students from various colleges involved in teaching Biotechnology and Microbiology.

On the whole, the conference was hailed by one and all as a memorable and fulfilling academic activity by the Department of Microbiology of Nagpur University (M.S.) India. *The Lab-to-Land programmes organized under this conference are also a tribute to the efforts of Rajiv Gandhi Vikas Biotechnology Centre, Nagpur.*

Several important recommendations have emerged out of the deliberations in the Global conference and the panel discussion (including open-house session). Some of the prominent and pertinent recommendations for emerging out of the delebration of the Global Biotech Congress-2000 *AD for consideration and implementation by the developing and developed countries of the world are as under:*

1. **Information Transfer Technology**

1.1 Lab-to-Land Technical session programmes should be conducted twice in a year at the villages, one each before the Rabi and Kharif seasons. Such programmes can be made more effective by using video-films and booklets specially prepared for this purpose in local languages. Validity Demonstration Trials should be laid at fields owned by the farmers with biofertilizers and biopesticides to convince the farmers of their efficacy. The activities should be coordinated by local experts having first-hand knowledge about the areas ensuring larger number of student participants, and experts including those from abroad. Various Government and Non-Government organisations should sponsor such activities, particularly those involving technology transfer.

1.2 Present day Students, who will have to carry on the future efforts towards sustainable development, would have to be activity involved ("hands-on" involvement) in using simple yet innovative biotechnological tools and in the practical demonstration of their utility.

1.3 The students, teachers and research personnel should be provided with facilities for on-line access to internet to keep them abreast of latest developments and to solve their experimental problems.

1.4 Industries, particularly Agro-industries should be encouraged to participate in such interactive programmes and interface ventures.

2. Sustainable Development

2.1 It is necessary to impart awareness that biotechnology is compatible with sustainable development. In fact, some research lines should be addressed directly to guard against environmental hazards.

2.2 The concept of sustainability is very much a part of the traditional wisdom of our rural communities. Biotechnology is a set of modern tools that can be utilized for achieving sustainable agricultural and economic development. Thus, bringing together the opinion leaders from the rural communities and biotechnological experts on a common platform at regular intervals, for effective cross-fertilisation of ideas and their implementation.

3. Biosafety

3.1 A broad legal framework should be established in developing countries like India to protect the important and indigenous biotechnology research findings under intellectual property rights.

3.2 Efforts should be made to encourage biotechnology research to control environmental pollution and to restore biodiversity.

3.3 It is imperative to explore the paradigm shifts in biotechnology and explore the outcome for improving the quality of life.

3.4 Biotechnology research should ensure sustainable consumption and sustainable development and reduce the prevailing uncertainties.

3.5 There is an urgent need to set up local germplasm banks comprising 3 to 4 district, to preserve biodiversity.

3.6 No Genetically Manipulated Organism (GMO) should be released along with any transgenic forms unless adequate safety measures are taken and unless acceptability by the local people, scientists and environmental activists is ascertained.

3.7 Possibility of 'R' factor among transgenic plants has to be ascertained before embarking on high-budgetory R and D projects.

3.8 After ensuring toxic levels, geographic agro-climatic conditions, pest infestation rating, safety factors etc., assurance of low-cost supply of transgenic seeds/ plants from R and D centre should be made mandatory. The competent authorities should ensure that the local germplasms are not lost, while using the transgenic forms.

3.9 Transgenic crops should be developed with improved agronomic traits such as herbicide tolerance, pest and disease-resistance, pollination control, delayed ripening, etc. and be made available for commercial cultivation in developing countries.

4. **Health**

4.1 The R and D efforts on Low-cost waste-treatment technologies should be encouraged in Public Health Sector.

4.2 Monitoring and detection of Plant and Human Pathogens by using molecular techniques is necessary for diagnosising various diseases.

4.3 There is urgent need to develop gene data bank in developing countries like India.

4.4 Greater emphasis is needed for the development of plant-based vaccines. In India, we should initially concentrate on the development of a few specific plant-based edible vaccines for testing the technology for commercial application.

4.5 Global level organisation has to be set-up for monitoring/regulating the implementation of the ban on dreaded human pathogens and chemicals which may be misused for biological and chemical warfare.

4.6 Proper regulatory procedures must be developed and adopted for eradication of epidemic forms of pathogens and their vectors.

4.7 Drug resistance in human pathogenic microorganisms is a major health hazard in the world today. To overcome this problem, use of indigenous herbal drugs should be encouraged. More and more research should be carried out in this area to prevent development of such resistance. In this context, drug regulatory legislation act should be urgently formulated and sincerely implemented.

4.8 Genetically modified banned biotech products should not be permitted for export or import to other countries. Similarly, the revolutionary biotech products not accepted in the developed countries by the local people shall not be exported to other developing countries. In such cases sternest measures need to be adopted by the Govts. of developing countries.

5. Environmental Bioremediation

5.1 Use of bio-fertilizers should be encouraged in remote villages for sustainable agriculture, and also for improving production unit of country through bioreclamation of Waste land/Barren land.

5.2 Bio-degradable materials should preferably be used for immobilization of cyanobacterial biofertilizers.

5.3 Multiple applications of biotechnology in paint industry may be explored with the help of R&D endeavour on the possible use of natural biocides in different paints, their synergic/antagonistic effects in various paint components, etc.

5.4 Application of specialized microorganisms for bioremediation of contaminated soils and other eco-systems is the need of the hour.

5.5 Phytoremediation is necessary with respect to metal contaminated eco-systems.

5.6 Role of tissue culture needs to be promoted to develop stress-tolerant plant species suited to contaminated soil.

6. R&D Funding/Human Resource Development

6.1 Science the 21st century is expected to have major thrust on biotechnology, it is necessary for the Government to start more biotechnology P.G. Centres and Centres of excellence in biotechnology research oriented towards specific applications.

6.2 The R&D efforts in transformation technologies should ensure cost-effectiveness and high reproducibility.

6.3 Further research is needed to elicit the mechanisms of producing secondary metabolites by plant hairy roots, action of elicitors, alteration of secondary metabolite pathways, modification of plant architecture, phytoremediation, etc. Scaling-up of special bio-reactors will go a long way to establish commercial applications of hairy root cultures.

6.4 Scientists should strive hard to develop cost-effective, eco-friendly and sustainable indigenous technologies.

6.5 The products developed by biotechnology should be commercialized through a designated Agency.

6.6 There is an urgent need to establish an extension wing for every Biotechnological Centre so that the insights and inventions of these institutions can trickel down to commenest of common people belonging to areas which are vulnerable to exploitation by capitalists, brokers and even the organized caucus of expertised related to biotechnological enterprises. There should be mechanisms for strict vigilance by Authorities to protect these ignorant people from exploitation and perpetual poverty.

Sincere efforts for implementing the above recommendations will go a long way towards the highly cherished goal of global sustainable development through biotechnology.